AF701937

Evaluating Residences Impacted by Smoke and Products of Combustion

A Practical Guide

By

Environmental Information Services, Inc.

Technical Handbook by J. Lieberman
Environmental Information Services, Inc.
4145 Piedra Court, Suite 100, Boulder, Colorado 80301
Web site: www.eis1.us E-Mail: Support@eis1.us

Please note that the majority of pictures in this handbook were taken by Mr. Lieberman, unless otherwise noted, in which case photographs are used with permission.

For permission requests regarding this manuscript, write or email the publisher, addressed "Attention: Permissions Coordinator" at the mailing address or email address below.

Environmental Information Services, Inc.
4145 Piedra Ct, Suite 100
Boulder, CO 80301
support@eis1.us
www.eis1.us

ISBN 978-0-9846825-3-9
Printed in the United States of America FIRST EDITION

Preface

The intent of this handbook is to provide readers with an in-depth understanding of the modern art and science of evaluating residential living spaces (single family homes, apartments, townhomes, condominiums and similar) that have sustained impacts from products of combustion (smoke, char, soot, and ash) and the associated contaminants likely to be deposited within and on the residence. This handbook will also strive to translate research, observations and the results of sampling and analysis into useful impact classifications that may be used to develop recommendations for remediation and restoration.

As a chemist and professional Certified Industrial Hygienist (CIH), James Lieberman, owner, founder, and president of Environmental Service Inc. (EIS) has conducted many post fire evaluations and recognizes that there is a dearth of information on a holistic approach to evaluating impacted residences. It is his hope that this guide will be useful for industrial hygienists, safety and occupational health professionals, remediation contractors, public adjusters, government officials and insurance company representatives.

There are currently no Federal or State regulations for conducting an assessment or evaluation of residences impacted by smoke and products of combustion. Mr. Lieberman hopes that the guidance presented in this handbook will help to educate practitioners and empower them to create clear and more logical evaluations that are supported with appropriate field observations, photographic documentation, and location samples. Mr. Lieberman also hopes that some of his recommendations will become good practices for conducting safe, consistent, and effective assessments.

Since this is an evolving area of study and investigation, Mr. Lieberman would like to extend an invitation to readers to contact EIS with questions and constructive comments. Your input will help EIS update and improve this handbook. We anticipate periodically issuing updated additions as new information is gathered.

Please email questions or comments to: **info@eis1.us**

Limitations and Exceptions

The profession of industrial hygiene is dedicated to the anticipation, recognition, evaluation, communication, and control of human and environmental stressors. The fields of industrial hygiene and environmental science use scientific methodology, but often requires additional professional judgment and experience in fully evaluating exposure hazards to determine potential risks. Therefore, this handbook does not replace the need for a well-informed, educated, and experienced investigator.

The information provided in this handbook was prepared in accordance with customary principles and practices in the fields of industrial hygiene, safety, and engineering, and is believed to be accurate. But the field is evolving quickly, therefore the findings obtained, and the recommendations presented in this handbook are based on current knowledge, which may be imperfect. This warranty is in lieu of all other warranties either expressed or implied. EIS is not responsible for the independent conclusions, opinions, or recommendations made by others based on the information presented in this handbook.

Acknowledgments

We would like to acknowledge and thank the following individuals for offering their time, knowledge, experience, and encouragement.

Mr. Don Archibald, a retired Certified Industrial Hygienist, and member of the of American Industrial Hygiene Association – Rocky Mountain Section (AIHA-RMS), provided a technical review of the draft version of the manuscript. Don provided suggestions for additional materials, constructive criticism of the draft, and provided insights on ways to clarify different industrial hygiene concepts. He also made suggestions for clarifying confusing language in some sections, besides identifying typos, and needed spelling corrections. We greatly appreciate the effort Don expended to help create a better handbook and hope to work with him to educate the IH community on the existence of this handbook.

Ms. Sarah R. Gordon, M.Sc., founder of The Argo Writing Studio, assisted with the professional organization of the book through proofreading, technical editing, and suggestions for how to organize and format the manuscript.

Ms. Linda Hendrix, General Manager for SKC-West, Rocky Mountains; an active member of AIHA-RMS; and a trusted vendor for occupational health & safety equipment, provided a technical review of the draft manuscript. She pointed out inconsistencies in the text, typos, redundancies and made suggestions for improving the text that were spot on. Her efforts and encouragement are appreciated.

Ms. Leanne Lestak, a knowledgeable geographic information systems (GIS) specialist, made suggestions for finding maps and information concerning past wildfire burn areas on the world wide web.

Mr. Ethan Priest helped type portions of the manuscript and provided suggestions and assistance with graphic design.

Mr. Thomas Sanborn, P.E., and VP of Forensic Analysis & Engineering Corp. He and Mr. Lieberman have collaborated on a number of fire and smoke related projects. He has shared his knowledge and insights concerning building construction; the effects of heat and fire on structures; and requirements for reconstruction. His input has helped us prepare better assessments and his review and comments on the book's manuscript were invaluable.

Mr. Robert Strode, a Certified Industrial Hygienist at Summit Exposure and Risk Sciences and a long-time friend and member of the AIHA-RMS, provided an in-depth technical review of the draft version of the manuscript.

Mr. Steve Thomas, a long-time friend whose own home was impacted by the Marshall Fire, was willing to share his experiences working with his insurance company; claims agent; and remediation and restoration contractors. His descriptions of all of the challenges and tasks that needed to be completed by a homeowner gave insight and empathy for others whose home has been impacted by smoke and products of combustion. He provided the confidential documents that he used to support his damage claim which included a summary of loss; explanation of building replacement costs; a spreadsheet tracking the location remediated; who the work was performed by; the description of the activity; and cost. These detailed documents allowed me to see the actual level of detail and effort that needed to be exerted to have a satisfactory outcome.

We are truly grateful for all who helped to make this handbook a reality. Their assistance helped make a better document.

*All content, grammatical mistakes, typographical and spelling errors within the book are the sole responsibility of the author.

Photographer Credit

We would like to thank Craig Toogood, John Schenne and Patrick Kramer for allowing the use of their photographs to illustrate dramatic scenes and important methods found in this handbook. All pictures not credited to the three above are either in the public domain or pictures taken by Mr. Lieberman.

Table of Contents

INTRODUCTION

Background

In 2012, Environmental Information Services Inc. (EIS) was retained to assist homeowners in Colorado Springs, Colorado with claims for damages to their homes as a result of smoke and products of combustion (S&PC) resulting from the June 23, 2012, Waldo Canyon fire. At the time there was no publication to assist an industrial hygienist such as the now existing American Industrial Hygiene Association's (AIHA) Technical Guide for Wildfire Impact Assessments for the OEHS Professional[1], published in 2018. More methods are now available to sample a building or residence for physical or chemical markers of S&PC than there were back in 2012.

After the 2018 AIHA publication became available, useful insights were added to the field's body of knowledge, but Mr. Lieberman felt the need for a guidance document that addresses more complex and advanced methods to help the practitioner perform more difficult and nuanced assessments and how to translate the results into actionable recommendations.

The Colorado Marshall Fire on December 30 of 2021 created a situation where several of Mr. Lieberman's friends and acquaintances needed assessment expertise. EIS worked hard to provide useful assessments of residences affected by fire, smoke, and products of combustion. Our reports were supplied to homeowners, engineers, public adjusters, and insurance companies. We reviewed many other investigators' reports and found that although some were excellent, way too many were inadequate which led to poor remediation projects. These experiences spurred Mr. Lieberman to write this handbook.

There is a real disconnect between what industrial hygienists and environmental health investigators are reporting, and how insurance companies, public adjusters and remediation contractors are interpreting results. This handbook therefore strives to create more objective requirements for the remediation contractors based on the evaluation results; and assigns minimum remediation requirements based on categories for impacted residences created and defined by EIS.

This handbook also spells out specific objective clearance requirements that the remediated residence must pass to be considered completely remediated and returned to a pre-fire or unaffected status (category). Mr. Lieberman has offered his suggestions, opinions, and recommendations, some of which are novel. We will endeavor to let the reader know when Mr. Lieberman's opinions are presented versus the industry practices.

The goal of this handbook is to shed light on advancements in subjects that have not been examined in detail. As a result, suggestions are included that expand the investigator's horizons. We hope that you find this guidance to be of assistance, and we endeavor to continue adding to it as the body of knowledge expands concerning the evaluation and assessment of impacts from smoke and other products of combustion.

Post Fire Events, Tasks & Parties

"In theory," investigators should be informed of the entire sequence of events from the time of the fire until the final clearance sampling is completed, and the remediation is finished. In actuality the investigator may only participate in pre or post assessments of the residence.

Investigators should also be aware of the potential conflicts of interest, and the importance of establishing objective standards to avoid disputes amongst parties. Furthermore, owners should review guidance that helps them to understand the entire process.

A few references are included to assist with these efforts and to help the investigator understand the challenges and difficulties that property owners will face. This handbook provides a relatively short overview of the subject in the next few paragraphs. For a more in-depth discussion of these issues as they impact the restoration of a residence, please refer to:

- *Guidelines for Fire and Smoke Damage Repair, Second Edition by Restoration Industry Associates*[2]
- *After the Fire or Disaster: Dealing with Your insurance Company*[3]
- *The Red Guide to Recovery: Resource Handbook for Disaster Survivors: Second Edition.*[4]

Recommended Tasks for Owners to Complete[5,6,7]

(See examples of summary of loss at the end of this section.)

1. File a Home Insurance Claim

- If your home has sustained a fire (whether it's a minor kitchen fire, major house fire, damage from a wildfire, or any other sort of fire damage), file a home insurance claim as quickly as possible.
- Filing quicker means you'll receive your claims payment quicker; and you can get back to your normal life sooner rather than later.
- Upon filing a fire damage claim, your insurance company will want to see any proof of the damaged home and belongings. They'll also send out their own adjuster to survey the damages and make a determination as to how much and when your claim may get paid out.

2. Document Losses

- Document all the damage that occurred. Make a written list and take photos and videos of your home and any personal belongings that are damaged or destroyed.
- Ideally the photographic documentation will include both a video walkthrough and detailed photographs of specific high value items taken before and after the event.
- These photos and videos are proof that the fire occurred and can be used later in the home insurance claims process.

- If you have receipts for anything that has been damaged or destroyed, make sure you get those immediately. Update or create a contents list to support the home insurance claim.
- Note areas and items that are not readily observable (e.g., interstitial spaces, ventilation system components, clothing, etc.)

3. Secure Your Property

- Secure the property to avoid theft and vandalism.
- Isolate damaged areas if possible.
- If there is a question concerning habitability contact a CIH, local building official, engineer or licensed building inspector.
- If the home is damaged to the point it becomes uninhabitable and you must temporarily move out, it is especially important to secure the property.

4. Meet with the Adjuster

- Meet with the adjuster and show the adjuster places and things that were damaged that they may not have noted.
- Document the meeting with the adjuster to make sure that you have reviewed all necessary information.
- While you're meeting with the adjuster, ask about coverage, policy maximums, deductibles, and other important limits to coverage.
- Take notes during your meeting. Without detailed notes documenting the process, you will be in a weaker position to demand agreed-upon repairs.

5. Verify Your Home Insurance Coverage

- Another important step to take *before* a fire is to have a baseline understanding of your insurance policy. Determine the insurance policy maximums and deductibles of coverage's A, C and D of your homeowner's insurance policy.

  Coverage A: This homeowner's insurance policy protects the physical dwelling structure of the home and helps pay for repairs to your home or pays to completely rebuild your home in the event of a home fire. Determining the value of rebuilding a home is difficult in this inflationary period. Your home value should include a surplus margin for changes in labor and material costs. Industry best practices would recommend a surplus margin or buffer of 20%. Be sure to update your coverage based on current economic conditions on an annual basis.

  At minimum, your dwelling coverage amount should be the exact value of your home.

  Coverage C: helps pay to repair or replace personal belongings damaged by the fire, helps protect things like electronics, clothes, furniture, and other household items. Personal property coverage should be at least the replacement value of everything you own in your home.

  Coverage D: helps pay for living expenses if the home is deemed uninhabitable, and you must temporarily move out while your home is being repaired or rebuilt. Loss of use coverage helps to pay for things like a hotel or equivalent rental property, storage facilities, gas, and food that you wouldn't have otherwise paid for if you were still able to live in your home. Loss of use coverage is usually up to 20% of your dwelling coverage.

6. Get Multiple Repair Estimates

- An adjuster may recommend a home repair company to repair or rebuild your home, but it's always a wise idea to get a few estimates from different contractors so you can compare their services.
- Prices to repair or rebuild may vary significantly. Be sure to get quality estimates and guaranteed work for the price.

7. Advance on the Claim

- If the fire damage was bad enough to temporarily force you out of your home, you may be facing exponential additional expenses you weren't expecting. Ask the home insurance company for an advance on your claim. It is unlikely that the insurance company will pay early; an advance can help ease the burden.
- Note – Most insurance companies will pay homeowners immediately upon the completion of an adjuster's initial report. However, an *initial report is not final*. Insurance companies expect homeowners and hired contractors to *update their claims as work progresses*. As insurance companies process updates to claims, they make additional payments to the homeowner or contractors – sometimes these payments can be paid within just a day or two.

United Policy Holder's Roadmap to Recovery:

We strongly recommend that homeowners visit United Policy (UP) Holder's website located at https://uphelp.org/about/ and review their *Roadmap to Recovery* program as soon as possible. Homeowners will gain a significantly greater understanding of how the various parties involved in the entire process play an important role in returning their home to a pre-fire condition.[8]

The reader will find the following and much more available on the UP site:

The *Disaster Recovery Handbook and Household Inventory Guide*: A compact book that captures the highlights of UP's guidance on getting a fair insurance claim settlement for damaged or destroyed property and recovering after a loss. Affectionately known as "the little yellow book," this tool has guided thousands of people through the process of navigating repairs, rebuilding, finances, and emotions after a disaster. Written by survivors for survivors, along with expert advice from trusted consumer advocates, construction, tax, and financial professionals.

The Roadmap to Recovery Organizer. A diary for keeping track of details, contacts, progress, and problems.

Claim tips and information relevant to specific types of damages/losses and events.

Parties to the Claim and Restoration Process

The following is a list of the most common parties and stakeholders to the claim and restoration process:

Property Owner
Tenant
Claims Agent
Insurance Company Adjuster
Public Adjuster
Professional Engineer
Investigator / CIH
Remediation Contractor
Restoration Contractor (s)
Subcontractor

Parties and Their Interest

The property may be occupied by the owner or a tenant. If so, the owner or their agent will start a claim with the insurance company which will then have an agent assigned to the case.

The homeowner may retain a public adjuster (PA), or the insurance company will provide an adjuster at no charge to the insured. A public adjuster who has no relationship with your insurance company may charge a flat fee or may charge a fee of up to fifteen percent of the insurance settlement for his or her services. The policyholder may retain a PA to help settle an insurance claim on his or her behalf. PAs are state licensed and trained insurance professionals. The use of an attorney is a potential alternative to the PA in extreme cases.

The claims adjuster will enter the process and gather information and details to find a fair settlement price based on their determination of what happened during the incident. Note that while a public adjuster can assist with the claims process, he or she cannot collect funds above those the insured are entitled to under your insurance policy.

It is in the interest of the policyholder to receive full and fair compensation on insurance claims as quickly as possible so they may have their home returned to a pre-fire condition. The initial settlement will not necessarily be the final settlement. The policyholder should keep very good records to determine the actual costs to remediate the residence.

An independent professional engineer may be retained as a third party to offer expert opinion on the fire's physical impact to the residence and the necessary requirements to repair the residence.

An independent professional investigator, usually a Certified Industrial Hygienist (CIH), may be retained as a third party to offer expert opinion on the residence's habitability, contamination, health impacts to the residents and the necessary requirements to remediate and repair to the original condition. Most insurance companies will pay for this service. The cost of such services is usually in the range of two thousand five hundred to six thousand dollars and is dependent on the complexity of the investigation.

The adjuster will provide a detailed remediation / restoration listing that includes the necessary work task and associated price to be paid for that level of effort per area of the home referred to as the

Explanation of Building Replacement Cost Benefits for Homeowner Policy, Coverage A. We will refer to the area, work task and payment list as the *restoration pricing*.

The remediation / restoration contractor will be contracted by the homeowner. The remediation / restoration contractor will use the insurance company's protocol pricing or create their own to determine the price for the repair. It is important to reconcile that price with the one that the insurance company is willing to pay to have a specific type of remediation task preformed.

The remediation contractor or restoration contractor will have a number of subcontractors to perform specific repairs. We use the terms remediation to refer to removal of contamination, and restoration to conduct repairs and normal maintenance such as painting and carpet replacement.

The Independent Investigator's Goals

If you are the independent environmental health investigator, your interest should be to conduct an assessment to identify and categorize likely physical and chemical contamination of the residence. The level of contamination will be used to determine impact categories to determine appropriate techniques for management of risks and to establish appropriate general remediation and restoration recommendations.

A seasoned investigator who prepares offsite ahead of time will be able to conduct a site visit and gain sufficient information from visual, sensory and odor information to make a preliminary impact category decision. This handbook provides a questionnaire and other tools that should be used in conjunction with the offsite data gathering and onsite visit to make the categorizing effort easier and more objective. The primary investigator's goal should be to **avoid conflicts and misunderstandings by clearly documenting their work; make decisions based on available data and their observations; and make recommendations independently.**

Further Reading

After the Fire or Disaster: Dealing with Your Insurance Company
https://www.nolo.com/legal-encyclopedia/dealing-with-insurance-after-disaster-29640.html

Guidelines For Reviewing Adjusters' and Contractors' Estimates
https://uphelp.org/claim-guidance-publications/guidelines-for-reviewing-adjusters-and-contractors-estimates/

Understanding the Insurance Claims Payment Process
https://www.iii.org/article/understanding-the-insurance-claims-payment-process

COLORADO—AFTER THE FIRE Your Step-by-Step Home Insurance Claim Recovery Guide
http://www.rmiia.org/downloads/AfterTheFire_StepByStep.pdf

Roadmap to Recovery program https://uphelp.org/recovery/

Examples Of Insurance Company Estimates & Summary of Loss

StateFarm®

State Farm
P.O. Box 106169
Atlanta, GA 30348-6169
Fax: 1-844-236-3646
statefarmfireclaims@statefarm.com

Structural Damage Claim Policy

This estimate is priced based on estimated market pricing for the cost of materials, labor, and other factors at the time of the loss.

Adjustments in market pricing and timing of the repairs may impact the final cost of covered repairs. Should you or the contractor you select have questions concerning our estimate, please contact us. If your contractor's estimate is higher than ours, you should contact us prior to beginning repairs. State Farm will work with you and your contractor to determine the actual and necessary cost of covered repairs at the time repairs will be completed, subject to policy terms, conditions and limits.

- We want you to receive quality repair work to restore the damages to your property.
- We will provide you with a detailed estimate of the scope of the damage and costs of repairs. Should the contractor you select have questions concerning our estimate, they should contact your claim representative directly.
- Depending upon the complexity of your repair, our estimate may or may not include an allowance for general contractor's overhead and profit. If you have questions regarding general contractor's overhead and profit and whether general contractor services are appropriate for your loss, please contact your claim representative before proceeding with repairs.
- There may be building codes, ordinances, laws, or regulations that affect the repairs of your property. These items may or may not be covered by your policy. Please contact your claim representative if you have any questions regarding coverage which may be available under your policy.
- State Farm® cannot authorize any contractor to proceed with work on your property. Repairs should proceed only with your authorization.
- State Farm does not guarantee the quality of the workmanship of any contractor or guarantee that the work will be accomplished within any specific time frame.
- It is understood that the contractor is hired by you, our insured, and that they work for you - not State Farm.

If you have any questions or need additional information regarding your claim, please contact your claim representative immediately.

Date: Page: 1

06-28X3-36H

State Farm

Building Estimate Summary Guide

This summary guide is based on a sample estimate and is provided for reference only. Please refer to the estimate for specifics of your claim.

State Farm Insurance

Insured:	Smith, Joe & Jane	Estimate:	00-0000-000
Property:	1 Main Street	Claim number:	00-0000-000
	Anywhere, IL 00000-0000	Policy Number:	00-00-0000-0
Type of Loss:	Other	Price List:	ILBL8F_MAR 13 Restoration/Service/ Remodel F = Factored In, D = Do Not Apply
Deductible:	$1,000.00		

Summary for Dwelling

Line Item Total [1]			5,953.10
Material Sales Tax	@	10.000% x 1,520.00	
Subtotal			6,105.10
General Contractor Overhead [2]	@	10.0% x 6,105.10	610.51
General Contractor Profit	@	10.0% x 6,105.10	
Replacement Cost Value (Including General Contractor Overhead and Profit [3]			7,326.12
Less Depreciation (Including Taxes) [4]			(832.50)
Less General Contractor Overhead & Profit on Recoverable & Non - recoverable Depreciation			(166.50)
Less Deductible [5]			
Net Actual Cash Value Payment [6]			

Maximum Additional Amounts Available If Incurred:

Total Line Item Depreciation (Including Taxes) [4]	832.50		
Less Non - recoverable Depreciation (Including Taxes) [7]			
Subtotal		312.50	
General Contractor O&P on Depreciation	166.50		
Less General Contractor O&P on Non - recoverable Depreciation			
Subtotal			
Total Maximum Additional Amounts Available If Incurred [8]			
Total Amount of Claim If Incurred [9]			

Claim Representative

ALL AMOUNTS PAYABLE ARE SUBJECT TO THE TERMS, CONDITIONS AND LIMITS OF YOUR POLICY.

1. **Line Item Total –** Total value of all line items in the estimate plus possible adjustments for *labor minimums. Labor Minimum* is to cover a certain minimum number of hours for drive-time, set up time and applicable administrative costs and repairs.
2. **General Contractor's Overhead and Profit –** General contractor's charge for coordinating your repairs.
3. **Replacement Cost Value (RCV) –** Estimated cost to repair or replace damaged property.
4. **Depreciation –** The decrease in the value of property over a period of time due to wear, tear, condition, and obsolescence. A portion or all of this amount may be eligible for replacement cost benefits.
5. **Deductible –** The insurer will pay for losses, up to the policy limits, in excess of your applicable deductible.
6. **Net Actual Cash Value Payment (ACV) –** The repair or replacement cost of the damaged part of the property less *depreciation* and *deductible*.
7. **Non Recoverable Depreciation –** *Depreciation* applied to items that are not eligible for replacement cost benefits.
8. **Total Maximum Additional Amount if Incurred –** Total amount of recoverable depreciation after actual repair or replacement of the property.
9. **Total Amount of Claim if Incurred –** Total amount of the claim, including *net actual cash value payment and total maximum additional amount available if incurred.*

1002989
Date: 8/2/2022 7:40 AM

139928.1 01-23-2013
Page: 2

State Farm

Summary of Loss

Named Insured: ______________________ **Claim Number:** ____________

Coverage A - Dwelling

Limit of Liability: $ 1,024,150 00

Description:		Repl Cost$:		Rec Dep$:		Non-Rec Dep$:		Amt Over Limit$:		Subtotal$:
State Farm Estimate	$	53,435.00	$	0.00	$	0.00	$	0.00	$	53,435.00
Total(s):	$	**53,435.00**	$	**0.00**	$	**0.00**	$	**0.00**	$	**53,435.00**

Comments:

Coverage A - Other Structures

Limit of Liability: $ 102,415 00

Description:		Repl Cost$:		Rec Dep$:		Non-Rec Dep$:		Amt Over Limit$:		Subtotal$:
State Farm Estimate	$	5,551.69	$	0.00	$	0.00	$	0.00	$	5,551.69
Total(s):	$	**5,551.69**	$	**0.00**	$	**0.00**	$	**0.00**	$	**5,551.69**

Comments:

Coverage B - Personal Property

Limit of Liability: $ 768,113 00

Description:		Repl Cost$:		Rec Dep$:		Non-Rec Dep$:		Amt Over Limit$:		Subtotal$:
Per Payment Tracker Worksheet Dated 8/2/2022	$	3,369.06	$	0.00	$	0.00	$	0.00	$	3,369.06
Flower Planter - $30.00	$	30.00	$	0.00	$	0.00	$	0.00	$	30.00
Contents Cleaning per HRS Colorado	$	7,567.26	$	0.00	$	0.00	$	0.00	$	7,567.26
Total(s):	$	**10,966.32**	$	**0.00**	$	**0.00**	$	**0.00**	$	**10,966.32**

Comments:

Coverage C - Loss of Use

Limit of Liability: $ Actual

Description:		Total:		Amt Over Limit$:		Subtotal$:
Incurred additional living expense	$	1,242.19	$	0.00	$	1,242.19
Increased Xcel Energy	$	77.35	$	0.00	$	77.35
Total(s):	$	**1,319.54**	$	**0.00**	$	**1,319.54**

Comments:

Additional Coverages:

Description:	Limit of Liab$:	Rep Cost$:	Rec Dep$:	Non-Rec Dep$:	Amt Over Limit$:	Subtotal$:

Trees, Shrubs & Landscaping	51,207.50	$	382.00	$	0.00	$	0.00	$	0.00	$ 382.00
Total(s):		$	382.00	$	0.00	$	0.00	$	0.00	$ 382.00

Comments:

Payments Made

Date:	Description:		Amount:
1/14/2022	Payment for Cleaning Home and Personal Property	$	25,658.59
1/28/2022	Payment for Additional Living Expense	$	1,242.19
3/17/2022	Payment for increased utilities and personal property	$	3,216.10
3/24/2022	Payment for Paint and Planter	$	7,505.38
6/2/2022	Payment for Cleaning of Dwelling and Rug Replacement	$	10,807.99
7/13/2022	Steve Smith Designs Invoice to Install replacement plants in backyard	$	382.00
Total(s):		$	48,812.25

Comments / Supplements

Thank you for choosing State Farm!

Subtotal: $	71,654.55
Less Deductible: $	4,651.00
Total Payable: $	67,003.55
Less Total Payments Made: $	48,812.25
Net Payment: $	**18,191.30**

Legend

- Repl Cost = Replacement Cost
- Rec Dep = Recoverable Depreciation
- Non Rec Dep = Non-Recoverable Depreciation
- Amt Over Limits = Amount Over Limits
- Limit of Liab = Limit of Liability

Signature *Date*

** Your policy may provide for additional payments on a replacement cost basis for the Recoverable Depreciation listed above for Cov. A and Cov. B. for your Building and Personal Property. Please refer to your policy and any endorsements for specific time limits and additional settlement provisions. Please contact your claim handler if you have any questions.*

THE HARTFORD

Understanding Your Estimate

This guide is intended to help explain specific terms. Please refer to your individual estimate for specific details pertaining to your claim.

Estimate Cover Sheet

Insured: Hartford Policyholder
Property: 1234 Mill Street
Center City, MO

Claim Rep: Mary Jones — Business: (987) 987-6543

Estimator: Sam Smith — Business: (123) 123-4567

Claim Number PP001234567 **Policy Number:** 12XX123456 **Type of Loss:** Wind Damage

Key

Claim Rep: The name and contact information of the primary claim adjuster handling coverage, contents and additional living expenses

Estimator: The contact information of the damages specialist assigned to prepare the structural damages estimate

Claim Number: The number assigned to your specific claim

Estimate Information

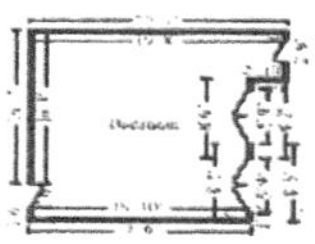

Bedroom — **Height: 8'**

633.33 SF Walls	349.19 SF Ceiling
982.53 SF Walls & Ceiling	349.19 SF Floor
38.80 SY Flooring	79.17 LF Floor Perimeter
79.17 LF Ceil. Perimeter	

	QUALITY	UNIT	TAX	RCV	AGE/LIFE	COND.	DEP%	DEPREC.	ACV
1. R&R 1/2" Drywall-hung, taped, floated, ready for paint	24.00 SF	2.13	0.94	52.06	5/150yrs.	Avg.	3.33%	(1.48)	50.58
2. Seal/prime then paint the walls (2 coats)	633.33 SF	0.79	8.25	508.58	5/150yrs.	Avg.	33.33%	(169.53)	339.05
3. Remove Carpet	349.19SF	0.21	0.00	73.33	5/10yrs.	Avg.	NA	(0.00)	73.33
4. Carpet	401.59SF	3.04	87.13	1,307.90	5/10yrs.	Avg.	50%	(653.96)	653.94
5. Drywall Labor minimum*	1.00 EA	275.72	0.00	275.72	0/NA	Avg.	0%	(0.00)	275.72
Totals: Labor Minimums Applied			0.00	275.72				0.00	275.72

Key

R&R: The **R**emoval and **R**eplacement of the damaged item(s) as covered by your policy

Unit: Cost of a singular unit of the repair item which includes all labor, material, and equipment to perform the repair

Replacement Cost Value (RCV): The total estimated dollar amount of your repairs

Age/Life: The reported age against the life expectancy of the item

Cond.: The condition of the item at the time of the loss, considering the life expectancy

Labor Minimum: An allowance made to complete all related repairs, including drive time and planning

Dwelling Summary — **Summary for Dwelling**

Line Item Total		2,824.95
Material Sales Tax		126.09
		$2,951.04
Less Depreciation		(979.71)
Actual Cash Value		**$1,971.33**
Less Deductible		(500.00)
Net Claim		**$1,471.33**
Total Depreciation	979.71	
Less Non-Recoverable Depreciation	<154.74>	
Total Recoverable Depreciation		824.97
Net Claim if Depreciation is Recovered		**$2,296.30**

Key

Depreciation (Recoverable) or <Non-Recoverable>: The loss of value, considering age/wear/tear/condition (**cond.**)

Deductible: The initial amount you are responsible to pay per policy declarations

Net Claim: The **RCV** less depreciation and less policy deductible

Net Claim if Depreciation is Recovered: The **RCV** less your deductible and less any applicable non-recoverable depreciation

Section 1.0 – Useful Lists

List Of Units & Formulas

SI Units

cc	cubic centimeter
cm	centimeter
cm^2	square centimeter
cm/s	centimeter per second
ft^2	square foot
ug/m3	microgram per cubic meter
m^2	square meter
mL	milliliter
mm	millimeter
mg/m^3	milligram per cubic meter

1 atmosphere = 14.7 psi or 760 mm Hg at standard temperature and pressure
1 liter = 1,000 cc or 1,000 ml
1 gallon = 3.785 liters
1 kilogram = 1,000 grams or 2.2 lbs

Useful Formulas:

1. To convert percent to ppm:
Contaminant value in percent x 10,000 = contaminant value in ppm

2. To convert ppm to percent:
$$\frac{\text{Contaminant value in ppm}}{10{,}000}$$

Scientific Notation

kilo = $1x10^3$
milli = $1x10^{-3}$
micro = $1x10^{-6}$
nano = $1x10^{-9}$
pico = $1x10^{-12}$

ppm = parts per million ($1x10^{-6}$)
ppb = parts per billion ($1x10^{-9}$)
ppt = parts per trillion ($1x10^{-12}$)
percentage (%) = parts per 100 ($1x10^{-2}$)

Corrosivity – pH scale

```
      0                                  7                                   14
Scale |----------------------------------|-----------------------------------|
      Acid                               neutral                             Base
```

List Of Abbreviations & Acronyms

ACGIH	American Conference of Governmental Industrial Hygienists
ASTM	American Society for Testing and Materials
ATSDR	Agency for Toxic Substances and Disease Registry
CIH	Certified Industrial Hygienist
COPC	Contaminant(s) of Potential Concern
EIS	Environmental Information Services, Inc.
EAA	Environmental Analysis Associates
EPA	U.S. Environmental Protection Agency
GC	Gas Chromatography
HPLC	High-Performance Liquid Chromatography
HVAC	Heating , Ventilation and Air Conditioning
IICRC	The Institute of Inspection, Cleaning and Restoration Certification
IPA	Isopropyl Alcohol
IRIS	Integrated Risk Information System
NIOSH	National Institute for Occupational Safety and Heath
OSHA	Occupational Safety and Health Agency
PA	Public Adjuster
PAHs	Polycyclic Aromatic Hydrocarbons
PC	Products of Combustion
PCB	Polychlorinated Biphenyl
PCM	Phase Contrast Microscopy
PEL	OSHA Permissible Exposure Limit
PPE	Personal Protective Equipment
REL	NIOSH Recommended Exposure Limit
RIA	Restoration Industry Associates
SDS	Safety Data Sheet
SVOC	Semi-Volatile Organic Compound
S&PC	Smoke and Products of Combustion
TLV	ACGIH Threshold Limit Value
TVOC	Total Volatile Organic Compounds
USDA	U.S. Department of Agriculture
VOC	Volatile Organic Compound
WTC	World Trade Center

List of Definitions and Description[9]

Aciniform Carbon is colloidal carbon having a morphology consisting of spheroidal primary particles (nodules) fused together in aggregates of colloidal dimension in a shape having grape-like clusters or open branch-like structures and have distinctively different origins. Carbon black is an engineered product.

Ash is pyrolyzed material obtained from advanced combustion of char. Because the carbon matrix of char is almost completely combusted, ash has a minerals-based matrix, with a high number of elements such as calcium, potassium, magnesium, aluminum, silicon, phosphorus, or sulfur. The particles are very brittle and may or may not contain some of the original material's cell structure.

Carbon black is a term usually associated with an industrial product, not soot, although they are different produced materials and should be used interchangeably. Carbon black is primarily composed of elemental carbon, obtained from the partial controlled combustion or thermal decomposition of hydrocarbons.

Char is defined in ASTM D6602-13 as particulate larger than 1 micrometer (μm) made by incomplete combustion which may not deagglomerate or disperse by ordinary techniques, may contain material, which is not black, and may contain some of the original material's cell structure, minerals, ash, cinders, and so forth. Carbon is typically the predominant component in char.

Corrosivity describes the acidic or basic quality of a substance to cause corrosion.

Soot is defined in ASTM D6602-13 as submicron black powder generally produced as an unwanted by-product of combustion or pyrolysis. It consists of various quantities of carbonaceous and inorganic solids in conjunction with *adsorbed and occluded organic tars and resins.* The EPA defines black carbon as the sooty black material emitted from gas and diesel engines, coal-fired power plants, and other sources that burn fossil fuel.

Soot/black carbon and carbon black terms should not be used interchangeably. The most important characteristics of soot/black carbon and carbon black particles are their size range and the aciniform morphology. Carbon black has as much higher surface-area-to-volume ratio than sot and significantly lower (PAH) content

List of EIS Impact Categories

The impact categories created by Environmental Information Services, Inc to define and describe groups of residential impacts from smoke, products of combustion and other contaminants are as follows:

- **Unaffected or Fully Remediated**
- **Mild**
- **Moderate**
- **Significant**
- **Severe**

List of Tables

Section 2.0 – Chemical Emissions of Fires

Fire Chemistry

A fire with a pure hydrocarbon fuel will result in nearly complete combustion. However, as more complex fuels burn, especially synthetic materials (carpets, furniture, personal items, etc.), the incomplete combustion generates many more complex products.

To further elucidate products of combustion, this handbook uses information found in Mr. Per Blomqvist's PhD dissertation "Emissions from Fires; Consequences for Human Safety and the Environment." Pertinent information from this dissertation is summarized below. [10,]

The composition of fire effluents includes fire gases such as carbon monoxide (CO) and hydrogen cyanide (HCN); irritants such as hydrogen fluoride (HF), hydrogen chloride (HCl) and isocyanates; and carcinogenic compounds such as benzene, polycyclic aromatic hydrocarbons PAHs, dioxins, and particulates.

Light-weight toxic smoke products include acid gases are produced from the combustion of halogen-containing-materials (where hydrogen chloride, HCl, is the most common); and a variety of organic compounds,[11] such as formaldehyde[12], acrolein and phenol. Heavier and more complex components include polycyclic aromatic hydrocarbons (PAHs)[13] and dioxins.

The effluents from a fire contain a mixture of gases, vapors, condensed compounds, and solid particles. The composition of the effluents is mainly dependent on:

- the elemental composition of the material involved in the fire
- the organic structure of the material
- the prevailing combustion conditions, i.e., temperature and ventilation

Wildfire December 30, 2021, moving from the West to the East over Marshall Mesa, Superior, CO

It is clear that the composition of the effluents varies strongly between fires and also during the course of a fire, but components that are found in almost all fire effluents are listed below:

- Carbon monoxide – CO
- Carbon dioxide – CO_2
- Benzene & other volatile organic compounds (VOCs)
- Polycyclic aromatic hydrocarbons – PAHs
- Soot particles of various sizes
- Ash
- Char

Secondary factors that influence the composition are the extent of dilution (cooling) of the effluents with fresh air (which quenches gas phase reactions); and the "age" of the effluents (which determines the time available for post-fire reactions). The latter point is particularly important for slow reactions, e.g., the development of particles (soot), and the deposition of condensable compounds.

Marshall/Superior Fire: Home in the Trail Ridge Neighborhood, Louisville, CO. Photograph provided by Patrick Kramer with permission.

It is important to remember that the composition of the effluents is not uniform and will vary strongly between wildfire and house fires, and also during the course of a fire. Components that are found in fire effluents, and which are associated with a certain risk, are schematically described in Table 2.0.

This handbook does not try to identify all of the potential chemicals that can be produced from burning a variety of different substrates. The investigator should do enough research to determine the likely chemical contaminants generated from the general types of materials consumed, and under which conditions the fire burned. It is particularly important to identify any unusual fuel sources that could have created a class of toxic compounds.

Table 2.1: Selected Components in the Fire Effluents and their Principal Associated Risks [14]

Type of Component	Examples of Compounds	Examples of Sources	Principle Risks
Inorganic gases	CO_2	All fires	Acute: asphyxia
	CO	All fires	Acute: asphyxia
	HCN	Nitrogen cont. Fuels, e.g., nylon	Acute: asphyxia
	NO, NO_2 (NO_x)	Nitrogen cont. Fuels	Acute: irritation Sublethal: lung damages
	NH_3	Nitrogen cont. Fuels	Acute: irritation Sublethal: lung damages
	HCI	Chlorine cont. Fuels, e.g., PVC	Acute: irritation Sublethal: lung damages
	HF	Fluorine cont. fuels, e.g., PTFE, PVDF	Acute: irritation Sublethal: lung damages
	HBr	Bromine cont. Fuels, e.g., Br-flame retarded material	Acute: irritation Sublethal: lung damages
	SO_2	Sulphur cont. Fuels, e.g., wool	Acute: irritation Sublethal: lung damages
Volatile organics	Isocyanates	Nitrogen cont. Fuels, e.g., mineral wool	Acute: irritation Sublethal: asthma, cancer
	Phenol	General for many fires	Acute: irritation
	Styrene	Polystyrene fires	Acute: irritation
	Benzene	General for all fires	Sublethal: cancer
Semi-volatile/condensed phase organics	PAH	General for all fires, particularly aromatic fuels	Sublethal: cancer
	Dioxins/furans	Fires with fuels containing chlorine or bromine	Sublethal: cancer, immuno-toxicity
Particles	Soot particles of various sizes	All fires	Acute: visual obscuration, Sublethal: deposition in the lungs

Contaminant Reservoirs

It would be highly unusual to have the investigator be in a position to sample a residence less than one week after a fire. If that were the case, one may want to include a variety of volatile organic compounds (VOCs) in their investigation, since they are present during and shortly after the fire. After a week or so, the concentration of any VOCs will diminish greatly, depending on the weather and ventilation of the residence.

The inorganic gases (such as carbon monoxide), oxides of nitrogen, and acid vapors (such as hydrochloric, hydrofluoric and hydrobromic acids) are usually produced in large quantities during active fires. Depending on many environmental conditions, some of the inorganic gas may react with moisture and form liquids, and others may be absorbed or adsorbed onto and into building materials. From EIS's experience, it is usually futile to try to sample for these compounds. Never-the-less, the investigation should understand that some materials will retain and re-emit vapors. We call these materials *contaminant reservoirs*, some of the most common contaminate reservoirs are carpet and padding, books, and upholstered furniture.

If an investigator enters into a residence and notes significant mucus membrane irritation (with onset of less than 15 minutes), it is usually caused by the out gassing of these types of acidic compounds, in addition to highly irritating VOCs. The irritation is a strong indicator of contamination and should be noted along with the time period that elapsed before it was initially noted.

If lingering VOC contamination is suspected, it is more productive to sample for VOCs after the remediation is complete using a TO-15 method[15]. EPA Method TO-15 is a procedure used to analyze air samples collected through canister sampling. "TO" stands for "toxic organics." It is used to measure up to 97 VOCs of the total 189 hazardous air pollutants listed in the Clean Air Act. It is used for testing grab samples of ambient air. These compounds can detect and are reported in a concentration as low as parts per billion (ppb), some in parts per trillion (ppt).

Usually, a significant period of time has elapsed between the time of the fire and the time that the residence is evaluated. The investigator will want to choose marker chemicals and physical particles that are likely to persist, and subsequently indicate the severity of the impact to the residence. Discussion of these chemical and physical materials follows.

Odiferous Markers

Many studies have described the smoky aroma characteristics of volatile phenolic compounds by gas chromatography-olfactometry methods and gas chromatography-mass spectrometry. There are different descriptive items for a chemical too, for example, 2-methoxy-4-methylphenol's aroma was described as phenolic, smoke, plastic, candy, spice and smoked, wood, smoke, and phenol. One problem with those studies is that they did not indicate the concentration of these chemicals. The odor character of compounds changes at different concentrations. Some researchers used highly trained sensory panelists to describe the sensory characteristics of several phenolic compounds associated with smoke aroma characteristics, such as p-cresol, m-cresol, thymol, guaiacol, 4-ethylguaiacol, eugenol, isoeugenol, carvacrol and 2,6-dimethoxyphenol. [16,17,18]

Most of phenolic compounds were described as smoky at low concentration, generally at 1 and 10 ppm, except 3,4-dimethylphenol at 5,000 ppm. This shows the powerful impact of phenolic compounds on smokiness. Smoky, woody, and dusty/musty attributes were found in the odor characteristics of most phenolic compounds. Among the 19 phenolic compounds, only 2,6-dimethylphenol was not associated with any smoky characteristics.[19,20]

The odor threshold of 2-methoxyphenol, or Guaiacol, was reported to be 0.001-0.002 ppm. Many halogenated substituted guaiacols were reported to have an odor threshold at parts per trillion concentration levels. Different studies have found the percent of isomers of methoxyphenols in wood smoke particulate matter to range from 0.1–6.5%. Although odor is one sign of contamination, it does not give the investigator a complete picture because the odor will change with concentration, and hazard and odor are not directly linked. [21,22,23]

Because many chemicals can be detected by odor at a concentration much lower than their exposure limit, *mucus membrane irritation combined with an odor is a much more reliable indicator of a problem.*

Physical Markers

The principal physical contaminants generated by fires include char, ash, and soot[24]; we refer to them as a group, as products of combustion (PC). Char is defined in ASTM D6602-13 as particulate larger than 1 µm made by incomplete combustion which may not deagglomerate or disperse by ordinary techniques that may contain material, which is not black, and may contain some of the original material's cell structure, minerals, ash, cinders, and so forth. Carbon is typically the predominant component in char. [25]

Char materials will always occur during fires as part of the natural process of burning. The mechanism of charring is part of the normal burning of certain solid fuels like wood. During normal combustion, volatile compounds created by charring are consumed in the flames or released to the atmosphere, while combustion of char can be seen as glowing red coals or embers which burn without the presence of flames.

Ash is pyrolyzed material obtained from advanced combustion of char. Because the carbon matrix is almost completely combusted, ash has a minerals-based matrix, with a high number of elements such as calcium, potassium, magnesium, aluminum, silicon, phosphorus, or sulfur. The particles are very brittle and may or may not contain some of the original material's cellular structure. Wood ash contains chemicals, which are very alkaline with a pH of 10 to 12. Thus, wood ash should be suspected to be alkaline, since about 80 to 90 percent of wood ash is water-soluble mineral matter. Electronic and other electrical equipment such as appliances can be adversely impacted by ash because of its alkalinity (corrosivity) and abrasive consistency.[26]

Soot is defined as submicron black powder generally produced as an unwanted by-product of combustion or pyrolysis. It consists of various quantities of carbonaceous and inorganic solids in conjunction with *adsorbed and occluded organic tars and resins.* A typical size of a soot nucleus is a few nanometers. The nucleus grows by additional condensation and coagulation. Freshly produced soot particles are almost spherical, and they tend to form, through coagulation, chainlike structures. Generally, fuels with higher carbon-to-hydrogen (C/H) ratio produce more soot. [27]

Chemical Markers

Polycyclic aromatic hydrocarbons (PAHs) are a group of organic compounds with relatively high boiling points that tend to be adsorbed on particles in condensed smoke vapors. These compounds tend to persist in the environment and may indicate whether residues from the event can still be detected.[28, 29,30]

PAHs are defined as being comprised of two or more joined aromatic rings. The toxicity of individual PAHs varies widely, and a range of non-cancer and cancer effects have been demonstrated. PAHs are semi-volatile/condensed phase organic compounds that are, to varying degrees, chemically stable compounds and lipophilic in nature. They are available for bioaccumulation and exhibit a range of toxic effects, of which carcinogenicity is the most significant. The group consists of a large number of individual compounds, for which the toxicity varies widely.[31,32]

Systems for toxicity ranking of PAH congeners have been proposed, in which *benzo[a]pyrene* (BaP) normally has a Toxic Equivalence Factors (TEFs) of 1.0. Under this system phenanthrene has a TEF of 0.001. The TEQ must be determined for the entire list of PAHs found in a sample to produce a relevant ranking of the hazard.[33,34]

As the smoke leaves the flames of a fire and cools, vapor phase PAHs condense on the surface of soot particles. The number of condensed organics varies from under 20% to up to 50%, for well-ventilated and under-ventilated fires, respectively. [35]

Structural fires generally produce orders of magnitude higher yields of PAH compared to residential combustion devices (ranges, fireplaces, candles, etc.). Room-fires were shown to produce a yield of approximately 1 g/kg of PAH. However, specific materials (e.g., flame retarded *high impact polystyrene* were shown to produce significantly higher yields, and a more toxic composition of the individual PAHs. [36,37]

Flame retardants are incorporated with polymers to improve flame resistance. Flame retardants act in numerous ways to provide fire resistance, for instance, by raising the ignition temperature; reducing the rate of burning; reducing flame spread; and reducing smoke generation. The normal flame retardants commonly used are brominated compounds.[38]

Dioxin & Furans

The world health organization considers dioxins an environmental pollutant. They are members of a group of dangerous chemicals known as persistent organic pollutants. Dioxins are of concern because of their highly toxic potential. Experiments have shown they affect a number of organs and systems. [39]

Once dioxins enter the body, they last a long time because of their chemical stability and their ability to be absorbed by fat tissue, where they are then stored in the body. Their half-life in the body is estimated to be seven to eleven years. In the environment, dioxins tend to biomagnify in the food chain. The higher an animal is in the food chain, the higher the concentration of dioxins. [40]

Until banned in 1979, polychlorinated biphenyls (PCBs) were manufactured as insulator fluids in heat-exchangers and transformers, as hydraulic fluids, and as additives to paints, oils, and caulks. Some of these chemicals remain in the environment even though they are no longer manufactured. They enter the food chain and are bioaccumulated and may become biomagnified. Burning PCB containing articles and oils create significant quantities of dioxins and furans.[41]

Dioxins, furans, and dioxin-like polychlorinated biphenyls are the abbreviated names for a family of chemicals that have similar toxicity and shared chemical characteristics. The dioxins and furans are no longer manufactured or produced intentionally but are created when other chemicals or products are made. These chemicals may be created during burning of forests, structural materials, and other household materials. Because of their hydrophobic nature and resistance towards metabolism, these chemicals persist and bioaccumulate in fatty tissues of animals and humans.[42]

The U.S. Environmental Protection Agency's (EPA) has developed an approach for evaluating the human health risks from exposures to environmental media containing dioxin-like compounds (DLCs). The chemicals 2,3,7,8-tetrachlorodibenzo-*p*-dioxin (TCDD) and DLCs are structurally and toxicologically related halogenated aromatic hydrocarbons. The EPA recommends that the toxicity equivalence factor (TEF) methodology, a component mixture method, be used to evaluate human health risks posed by these mixtures, using TCDD as the index chemical. The EPA recommends the use of the consensus TEF values for TCDD and the DLCs published in 2005 by the World Health Organization (WHO).[43]

Chronic exposure of animals to dioxins has resulted in several types of cancer. TCDD was evaluated by the WHO's International Agency for Research on Cancer (IARC) in 1997 and 2012. Based on animal data and on human epidemiology data, TCDD was classified by IARC as a "known human carcinogen". However, TCDD does not affect genetic material and there is a level of exposure below which cancer risk would be negligible.[44]

Due to the omnipresence of dioxins, all people have background exposure and a certain level of dioxins in the body, leading to the so-called body burden. Current normal background exposure is not expected to affect human health on average. However, due to the high toxic potential of this class of compounds, efforts need to be undertaken to reduce current background exposure. [45]

The joint Food and Agriculture Organization of the United Nations (FAO) and WHO Expert Committee on Food Additives established a provisional tolerable monthly intake of 70 picogram/kg (pg/kg) per month. This level is the amount of dioxins that can be ingested over lifetime without detectable health effects.[46]

Metals

Heavy metals may be a concern depending on the materials consumed in a fire. The term heavy metal refers to any metallic chemical element that has a relatively high density and is toxic or poisonous at low concentrations. Examples of heavy metals include mercury, cadmium, arsenic, chromium, thallium, and lead. [47]

As trace elements, some heavy metals such as copper, selenium and zinc are essential to maintain the metabolism of the human body. However, at higher concentrations they can lead to poisoning. Heavy metals are dangerous because they tend to bioaccumulate; which means an increase in the concentration of a chemical in a biological organism over time, compared to the chemical's concentration in the environment. Compounds accumulate in living things any time they are taken up and stored faster than they are broken down (metabolized) or excreted. [48]

Heavy metals can cause gastrointestinal and kidney dysfunction, nervous system disorders, skin lesions, vascular damage, immune system dysfunction, birth defects, and cancer. Simultaneous exposure to two or more metals may have cumulative effects. High-dose heavy metals exposure, particularly mercury and lead, may induce severe complications such as abdominal colic pain, bloody diarrhea, and kidney failure.[49] One should also note that children are much more susceptible to neurological function/developmental issues from heavy metals.[50]

Section 3.0 – The Investigation

Investigator Qualifications and Responsibilities

Qualifications

The investigator performing the assessment of residences impacted by smoke and products of combustion should have the education, training, knowledge, and field experience that would enable them to produce a satisfactory evaluation and report. The task also requires the investigator to be able to conduct a forensic investigation to establish contaminant origin, especially when the residence is outside of the burn zone. We believe that a suitable candidate would be a Certified Industrial Hygienist (CIH) that has mentored with an experienced S&PC investigator and also has knowledge of home construction. Industrial hygiene is the science and art devoted to the anticipation, recognition, evaluation, and control of those environmental factors or stresses arising in or from the workplace that may cause sickness, impaired health and well-being, or significant discomfort among workers or among the citizens of the community. A CIH with the above experience will have the education, expertise, and training to conduct a thorough investigation.

Responsibilities

The investigator should have access to specialized experts to consult with such as forensic lab managers, professional engineers, and restoration contractors. The investigator should communicate when additional work to further assess and characterize the extent of impact may be required after the initial assessment, especially if they suspect hidden building damage.

The evaluation and assessment report produced by the investigator must provide the necessary information to categorize the extent of impact and provide the basis for the adjusters, and remediation and restoration contractors to create a comprehensive remediation plan.

The investigators report should produce an opinion on the impact category. The impact category may then be used with the remediation and restoration matrix to provide guidance on the minimum cleaning and restoration actions to take, and what materials may be reasonably cleaned or should be disposed.

The investigator's report must also specify the fire origin, characteristics and how it produced the impact to the residence to satisfy insurance company requirements.

Required Elements of The Report

The assessment report identifies the type of the fire (either exterior wildfire or interior house fire); the associated combustion material such as natural organic materials, manmade materials, or a combination of both; the location of the fire; the expected characteristics of the smoke produced; and the layout of the residence. The report should describe the nature and location of the contamination found (both chemical and physical), visual observations, and textural characteristics such as oily soot and odor characteristics.

The characterization of contamination should include an opinion from the investigation whether the smoke resulted from oxygen-rich fires (dry) that burned fuel more efficiently and completely, or smoke from slower burning oxygen-starved fires (wet) that generated and deposited residues more evenly as a dark sticky film appearing as a stain, particularly on horizontal surfaces.

Collecting Off-Site Background Information

With larger wildfires, information such as the location, extent, weather, wind direction, and other conditions on the day or period of the active fire should be obtained through a search of the World Wide Web. A particularly valuable website for information is Wildfire Today (wildfiretoday.com), but there are other governmental and private sources available to the investigator such as the National Interagency Fire Center at https://www.nifc.gov/ and United States Geologic Survey's (USGS) Earth Resources Observation and Science (EROS) Center and Fire Information for Resource Management System (FIRMS).

Use the Inspection Questionnaire and Checklist to record your findings (found in Appendix B.)

The information available may include the type of fire (such as grass, forest or structural, or mixed) and reports concerning the winds direction and speed during the event. With larger fires, some sampling may have already been conducted. The results of such investigations may prove to be extremely valuable, and it is worth the effort to find out if such information is available. Check with local health and fire departments in addition to local newspapers. Refer to the excerpts from Marshall Fire Operations After-Action Report in Appendix A; for the full report see:
https://www.superiorcolorado.gov/home/showdocument?id=20708&t=637908398510628002

Please review the map of the Marshall fire below. Maps like this one can assist the investigator in locating the subject property's proximity to the fire.

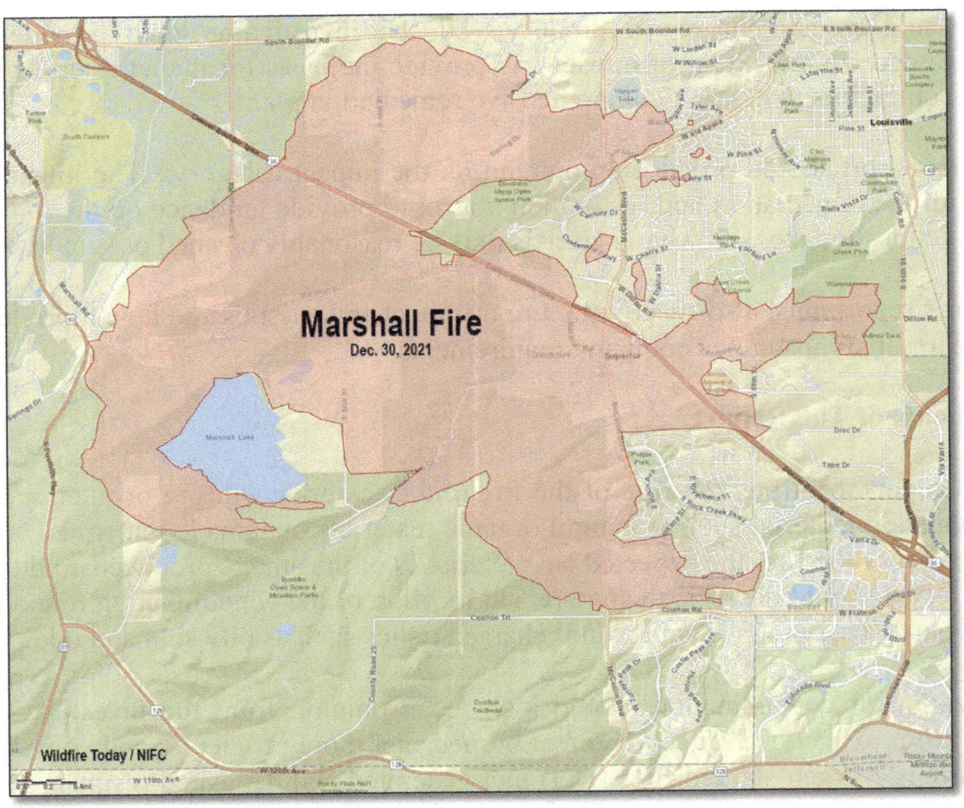

Collecting Onsite Interviews & Observations

Assessment of the subject property is not just about collecting samples!

Choosing the appropriate physical and chemical markers for sampling to assist in categorizing the contamination impact level of a residence must necessarily come after a thorough off-site and initial on-site inspection and assessment.

Before any decision is made on sampling, one should have completed the Inspection Questionnaire & Checklist (found in Appendix B). The form has sections that can be completed off-site and additional questions and tasks to be completed onsite. The form has questions about the location, nature, type, and proximity of the subject residence to the fire; if facilities that contained hazardous substances involved in the fire were nearby; if structures burnt to the ground; and if automobiles were incinerated. Parts of the questionnaire should be completed before the onsite visit, and it should be used as a tool during the initial on-site inspection.

The investigator needs to take a systematic approach in creating an accurate, credible, and documented assessment of impacts due to fire, S&PC and other potential contaminates. Once all the data requested by the questionnaire has been inputted (from the off-site research, on-site visual and olfactory investigation), the investigator will be in a much better position to determine the types and number of samples to collect. We provide recommended sampling guidelines in this handbook, but the investigator should use their own knowledge and experience to modify the suggested number and types of samples. *We strongly recommend collecting at least the minimum recommended number and types of samples.*

Once the off-site data gathering is complete, or concurrent with the review, the investigator should interview the building owner or residence occupants to gather additional information. It is prudent to conduct a phone interview to gather initial information; make sure that you are speaking with an individual knowledgeable about the specific site and determine who may arrange for an onsite visit. Once onsite, the initial interview should be expanded depending on the observations made during the onsite inspection.

The inspection questionnaire will assist the investigator in covering all of the important topics below:

- Information Gathering & Occupant Interviews
- Visual Inspection
- Odor Detection

The investigator should now be able to form a hypothesis for what potential category the subject property should fit in, ranging from unaffected to severely impacted. *Further investigation and sampling may be used to test the hypothesis to ensure that the investigation has made a well-informed decision, but the effort should be sequential and used to narrow down the possibilities.*

Categories of Impacted Residences

1. Unaffected/Remediated
2. Mild
3. Moderate
4. Significant
5. Severe

These categories generally correlate to the following definitions used in *Guidelines for Fire and Smoke Damage Repair*[51] *(Guidelines D&R)*, but they are not a direct fit:

Light Damage - Mild
Moderate Damage - Moderate
Severe Damage - Significant
Very Severe Damage – Severe

Note: *There is no category equivalent for unaffected/ remediated in Guidelines for D&R.*

<u>Description of Categories</u>

Unaffected/Remediated

The unaffected or remediated residence *will not have* visually observable or olfactory-sensed evidence of any impacts from S&PC; nor will any sample of the residence exceed applicable background or screening levels for contaminants. The residence will look, feel, and smell like a normal home. It will either be or have been restored to a pre-fire impacted condition.

Mild

The mildly impacted residence will normally have visually observable and olfactory-sensed evidence of impacts from S&PC. Samples collected from the residence may exceed applicable background or screening levels for physical PC contaminants (such as char, ash or soot), but the impacts will be limited to those that could normally be removed by cleaning activities. The residence is not in a pre-fire condition and will need some minimum level of remediation to be returned to such condition.

Moderate

The moderately impacted residence will have visually observable and olfactory-sensed evidence of impacts from S&PC (e.g., the ash and char). Samples collected from the residence will exceed applicable background or screening levels for physical PC contaminants (such as char, ash, or soot), but not necessarily chemical markers. The impacts may necessitate professional cleaning / remediation and may require furnishings to be specially cleaned or disposed of before the residence can be returned to a pre-fire condition.

Significant

The significantly impacted residence will have unmistakable and wide-spread visually observable and olfactory-sensed evidence of impacts from S&PC. Samples collected from the residence will exceed applicable background or screening levels for physical PC contaminants (such as char, ash, or soot), and most likely for chemical markers of contamination. The residence will not be habitable until remediation & restoration are complete. The impacts will necessitate professional cleaning / remediation and may require furnishings to be specially cleaned or disposed of before the residence can be returned to a pre-fire condition. Walls will need to be sealed and painted, carpet & padding disposed, and other more aggressive remediation and restoration activities may be required.

PC on windowsill.

Heavy PC deposited on sliding glass door track.

PC & other debris in garage.

PC debris in garage entry area.

Severe

In addition to the same characteristics as the significantly impacted residence, the severely impacted residence will have sustained actual damage to the structure and components from heat or the fire. The residence will not be habitable until remediation & restoration. This residence will not only require professional cleaning / remediation, but also some demolition and reconstruction for restoration to be completed. The scope of the restoration will require the services of a Professional Engineer (PE) to assess the structural damage to the residence. A PE will need to determine the extent of demolition and replacement of windows, doors, dry wall, plumbing, electrical and other items before reconstruction.

Burnt deck 4" x 4" support & retaining wall

"Ghosting" on bathroom floor.

Selecting Preliminary Categories

In addition to the descriptive categories for impacted residences, this handbook also describes the types and extent of physical and chemical contamination associated with each category described previously in this chapter.

To create appropriate general remediation and restoration recommendations, and to determine appropriate techniques for management of risks, this handbook also adapts banding as embodied in the previously described categories as a technique to guide the practitioners' assessment, identification, and categorization of impacts. A seasoned investigator who has done their offsite work will be able to conduct a site visit and gain sufficient information from visual and odor information to make a preliminary decision. The questionnaire designed for this handbook, along with other tools used in conjunction with the onsite visit, should be utilized to make the categorizing effort easier and more objective.

This approach is based on two pillars: the fact that there are a limited number of identifiable impacts; and identifying the general severity allows the investigator to select approaches and techniques for further sampling and analysis to confirm or negate that the investigator's initial assessment was accurate.

It is important to note that banding by categorization is NOT a replacement for experts conducting field assessments nor does it eliminate the need to collect samples to support the practitioner's observations and assumptions in most instances. The use of the questionnaire and applications of intelligent data gathering methods will help to create more of a consistent and systematic approach to investigations.

Impact Categories: Physical Characteristics

Unaffected/Remediated:

1. No physical damage to the building, lot, and/or structure from fire or S&PC.
2. No signs of staining or heat damage to the exterior or interior of the structure.
3. No smoke odor sensed within the residence.
4. Ash, char, and soot (PC) as identified using tape lift sampling are at background levels adopted from EAA as less than or equal to 3% area ratio percent of tape slide or surface density measured of less than 5 counts per square millimeter (cts/mm^2).

Mild:

1. No structural damage from the fire to the interior or exterior of the building.
2. Interior and exterior show no signs of damage from heat or staining from soot.
3. Smoke odor may be present inside, but it does not create significant mucus membrane irritation (no signs of irritation after 30-minute exposure).
4. S&PC, if present, are visually deposited only at entryways and unsealed openings of the structure.
5. Ash, char, and soot (PC) on vertical surfaces (walls) or ceilings, as identified using tape lift sampling, may exceed background, but are localized within a 10-ft radius from entryways and has not affected more than 10% of interior surfaces of the residence as measured by floor square footage.

Moderate:

1. No structural damage from the fire to the interior or exterior of the structure.
2. Interior and exterior show no signs of damage from heat but may have some staining from soot.
3. Smoke odor is present inside and may create mucus membrane irritation within a 15-to-30-minute exposure.
4. S&PC are visually deposited at entryways and unsealed openings of the structure in addition to other locations.

5. Ash, char, and soot (PC) on vertical surfaces (walls) or ceilings, as identified using tape lift sampling, may exceed background, and may have impacted 10% of the residence but less than 20 percent as measured by floor square footage.

Significant:

1. Fire may have damaged nonstructural portions of the building and surrounding areas such as attached decks or perimeter fences.
2. Interior or exterior may show signs of heat damage such as blistering, or stained paint, or warping of windows.
3. Smoke odor is heavy and may be irritating with exposures of less than 15 minutes.
4. S&PC are visually deposited at entryways and unsealed openings of the building.
5. Ash, char, and soot (PC) on vertical surfaces (walls) or ceilings, as identified using tape lift sampling, exceed background and has affected more than 20% of the residence as measured by floor square footage. S&PC are greater than 10% area ratio percent of tape slide or surface density measured is more than 20 counts per square millimeter (cts/mm^2).

Severe:

1. Fire will have damaged structural or nonstructural portions of the building and surrounding areas such as attached decks or perimeter fences.
2. Interior or exterior may show signs of heat damage such as blistering, or stained paint, or warping of windows.
3. Smoke odor is heavy and may be irritating with exposures of less than 5 minutes.
4. S&PC are visually deposited at entryways and unsealed openings of the building.
5. Ash, char, and soot (PC) on vertical surfaces (walls) or ceilings, as identified using tape lift sampling, exceed background and has affected more than 20 percent of the residence as measured by square footage. S&PC are greater than 10% area ratio percent of tape slide or surface density measured is more than 20 counts per square millimeter (cts/mm^2).

Upon completion of the questionnaire and initial onsite investigation, it may be appropriate to collect samples. The initial observational categorization should help the investigator select the appropriate sampling methods for the types of potential contaminants to identify. Refer to guidelines for sampling according to impact category in Section 4.0 of this handbook (*note - a written sampling plan must be developed before conducting sampling.)* Some of the types of contaminants that may be sampled for are:

- PCs (Char, Soot, Ash)
- Debris in carpet
- VOC's
- pH - Corrosivity
- SVOC's
- PAH
- Heavy Metals
- Other

When Investigation & Observations Are Sufficient

Sometimes the source of the fire and contamination are so obvious that it only requires the investigator to do background research on the contaminants likely generated, and complete visual observation to support a conclusion that identifies an impact category. (You may, by analogy, use it to identify others that fit the situation where the source of the contamination is known, and the observational information is extremely clear.)

Please refer to pages 76-77 for pictures of an iPad fire that occurred in a residential kitchen. These pictures are taken from a residence that had an iPad spontaneously combust as a result of the lithium battery igniting. It was in a plastic vinyl case located on a wood kitchen table. The resulting smoldering fire generated a significant amount of smoke that deposited an oily soot on walls and ceilings. The kitchen was located on the ground floor level of the home; two bedrooms and a bathroom were located on the second floor. The home had a basement in which approximately 50% was finished. Please note that the author was called in to assist after significant but unsuccessful remediation activities had been conducted.

Let's examine the available and observed evidence that was used to determine the category that this home should have first been placed.

The residence had widespread soot on walls, ceilings, and in the gas forced air HVAC system. The HVAC ducts were coated with an oily soot. Immediately after the fire, there was a strong odor of smoke, and the interior atmosphere was irritating to mucus membranes. According to the available literature, smoldering fires, especially from electrical equipment, generate toxic smoke.

We already know from research that plastic fires have the capability to generate VOCs, PAH, dioxins, and other toxic and irritating compounds. These compounds contribute to the toxic/irritant quality of smoke gas atmospheres. Lithium battery fires generate a whole host of irritating and toxic chemicals.

One could argue – from observational information, knowing the source of the fire, and from conducting research on the likely products of combustion from the source – that the residence was significantly impacted from the fire. Without collection of a single sample, an investigator could preliminarily categorize this impact as significant; furthermore, one could *also say that the impact has some characteristics of severe* since the residence sustained actual damage to components from heat or the fire. The residence was not habitable until remediation & restoration were complete.

Sampling would be valuable as an aid to determine the extent of contamination, but not the impact category for this residence.

Examples of Residence Layout Sketches

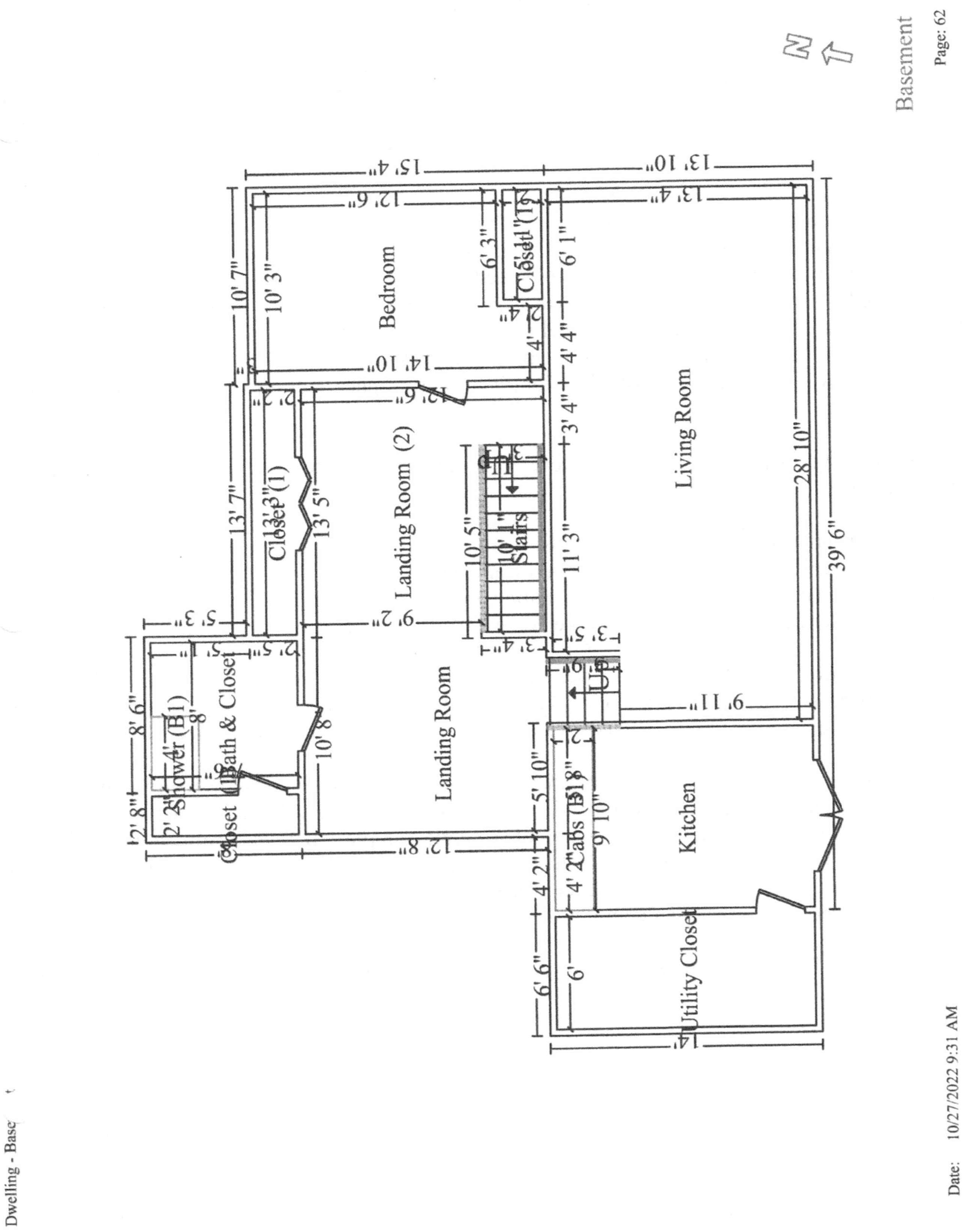

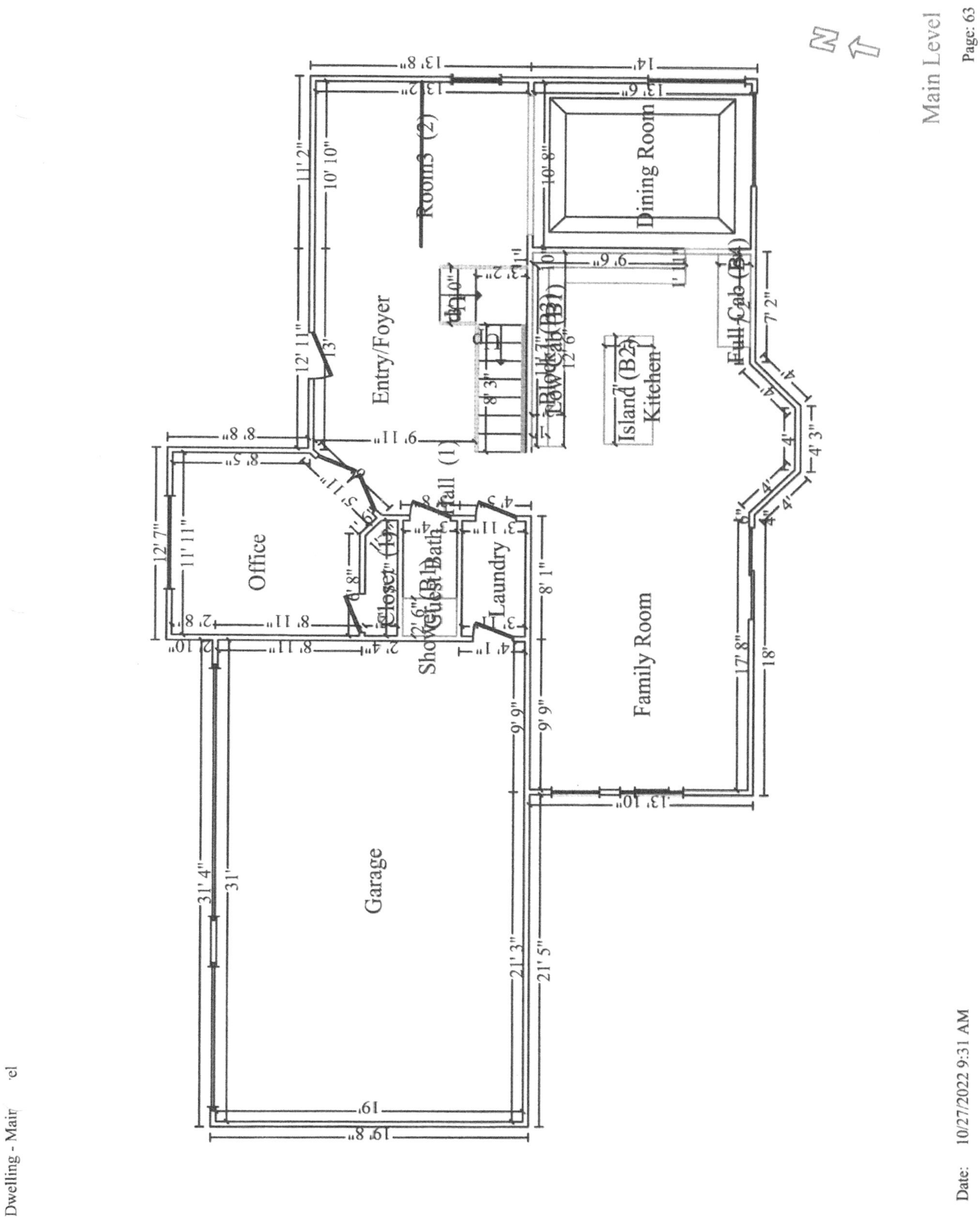
Main Level
Page: 63
Room3 (2)
Dining Room
Entry/Foyer
Kitchen
Island (B2)
Office
Closet
Guest Bath
Laundry
Hall (1)
Family Room
Garage
Date: 10/27/2022 9:31 AM

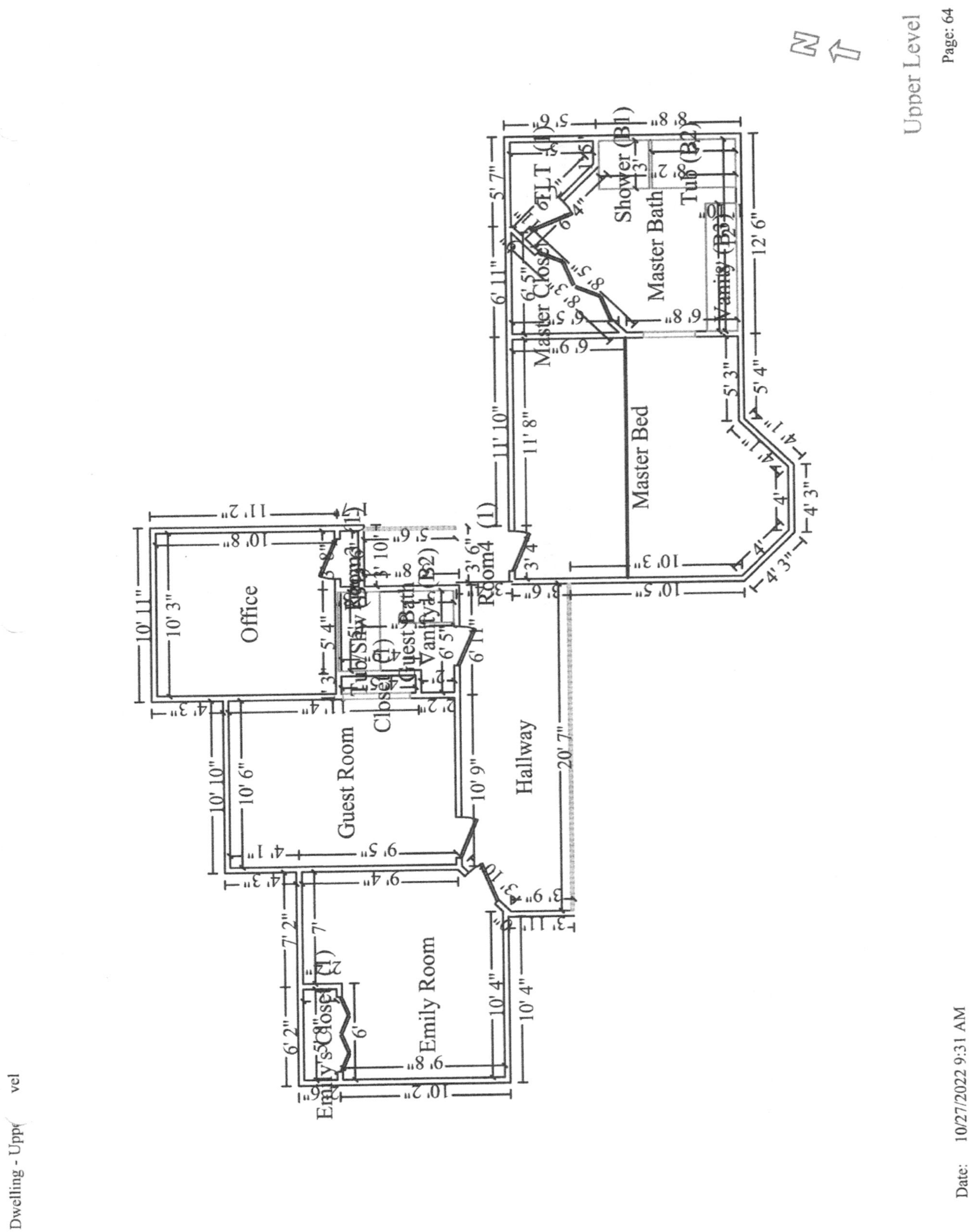
Upper Level
Page: 64
Office
Guest Room
Hallway
Emily Room
Master Bed
Master Bath
Date: 10/27/2022 9:31 AM

Section 4.0 – Sampling Logic, Methods & Materials

Sampling Considerations

Sampling is a vital component of the investigator's ability to assess the impact from smoke, other products of combustion and associated contaminants. The purpose of sampling is to identify the fire source and type; support the categorization decision; determine the extent of contamination; identify the contaminants that exceed screening levels; and provide the bases for remediation and restoration.

Key Recommendations

- *Think & plan before collecting any samples!*
- *Never collect samples unless the results are intended to either prove or disprove a hypothesis.*
- *Never collect a sample unless you have a comparison scale to evaluate the results against! A number without a reference value is worthless.*
- *If you are going to sample, collect a sufficient number of samples to be statistically relevant; <u>this usually requires a minimum of three samples.</u>*
- *You should be able to explain clearly what the results mean.*
- *You should be able to defend your sampling plan, method and results interpretation as an expert if asked to testify in a court of law.*
- *Samples that are not well documented are of no value!*

Sampling Periods and Timeline

Sampling may be conducted during the initial onsite investigation or soon afterwards to support the categorization of the impact to the residence; we refer to this as categorization sampling. When the source of the impact is not visually obvious, the investigator should make sure that the supporting laboratory and microscopic analysis data can help document the origin of the fire. Samples may also be conducted during the remediation process to ensure that the areas cleaned by the contractor actually meet the established contamination reduction criteria or acceptable residual levels; we refer to this as check sampling. Once the remediation contractor has stated that the scope of work contracted for has been completed and the home should now meet agreed upon contamination levels, an additional type of sampling is final post-remediation sampling; we refer to this as clearance sampling.

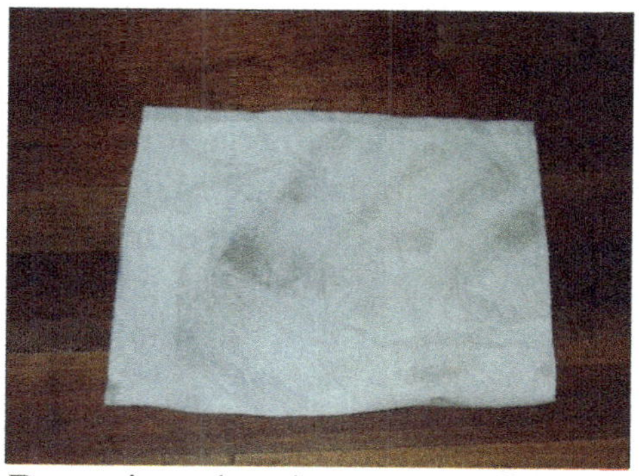

Dry wipe showing soot collected. from a wall.

This handbook provides a recommended set of criteria for allowable residual contamination levels. If there is no sampling and analysis information on the residual contamination levels in the home post-remediation, then there will not be a way to be sure that it has been cleaned adequately nor will information exist to enforce the remediation contract. The homeowner will have to assume that the job has been faithfully completed. Therefore, we always recommend clearance sampling, see Evaluation Criteria for Remediated Homes in Appendix C.

Use of Clearance or Check Sampling

Sampling may produce results that assist the investigator with the pre-remediation categorization determination of the impact level, but post-remediation sampling is very important to ensure that the residence is truly clean and has been returned to a pre-fire unaffected state.

Check sampling is another type of sampling that is especially important but seldom conducted. Check sampling is used to determine the effectiveness of an ongoing remediation. We suggest that check sampling be employed if one suspects that the remediation contactor is not meeting expectations to *ensure that the remediation contractor is truly producing a quality job.*

Aggressive air sampling.

Micro-vacuum sampling.

We prefer to conduct a visual inspection and collect tape lift, wipe or micro-vacuuming samples, or a combination thereof, to check the quality of the remediation job *prior to final payment* made to the remediation contractor. We also recommend similar sampling methods for clearance sampling, with the addition of aggressive air monitoring. These tests can be completed relatively quickly so an answer can be obtained before the contractor demobilizes. We define aggressive air monitoring as employing a fan with an output of 2,000-2,500 cfm to circulate air in a room in which air monitoring is being conducted for char, ash and soot, or mold spores.

Objective

The investigator must always have a stated or implicit hypothesis to test; using the results of sampling and analysis to validate or contradict the hypothesis. This manual will provide the logic, tools, and suitable methodologies we employ to collect samples that will provide useful data for determining the impact category for the subject property. Furthermore, we will endeavor to describe the sample logic, sample selection, collection of samples, how to evaluate results and usefulness of the data for making intelligent categorization of the impacts to a subject residence.

General Guidelines

Strive to identify and measure specific fire-generated marker chemicals that would normally be detected if residual chemical contamination was present in the home. Compare the concentrations of the chemical markers to established guidelines or screening levels to determine if they are above screening or background levels and thus may be creating a hazard to human health or the environment.

A fully remediated residence should be returned to an essentially pre-fire condition; the category we describe as Unaffected/Remediated. In most cases this means that contamination levels should be equal to or less than background or established screening levels. In some cases, levels will need to be lower, for instance if the residence was home to a particularly sensitive person and the homeowner had taken steps to ensure that the home was habitable for that individual, the remediated home should be returned to the same equivalent condition using appropriate sampling method(s) for the agents of concern.

It is extremely important to identify the goals and objectives of the planned sampling event. The goals and objectives will differ significantly depending on if they will be used to assist in the initial categorization of the impact to the residence, remediation progress or post remediation clearance sampling.

A written sampling plan is absolutely essential. The plan needs to identify what real-time monitoring may be conducted, where samples will be collected, the type of samples to collect, and numbers of samples. This allows for the acquisition of the right equipment and supplies to support the effort, and coordination with the laboratory for needed supplies or solvents. The plan may be modified during the actual sample collection, but if one does not have a written plan, then it is much more likely that important samples will not be collected. The goals of the sampling must be clear, thought out and reflected in the plan.

When conducting initial or pre-remediation sampling of the residence, similar efforts conducted for nearby residences will prove to be very valuable and may help to target the types of samples to collect. Before any sampling event, the investigator should review and digest the information obtained from the impact questionnaire to form a hypothesis as to the impact category and form an opinion as to the type and number of samples needed. This effort will allow the investigator to translate the information into a sampling plan. The sampling plan should state the type of samples to be collected, the method and the number of samples. Refer to Guidelines for Sampling According to Presumed Impact Category later in this section.

Sampling data should be used to refute or support the investigator's hypothesis concerning contamination types and concentrations. For example, if automobiles were incinerated near a residence, the investigator may want to collect wipe samples to investigate the presence and concentration of metals. The hypothesis is that the vehicle contained a lead acid battery and other components containing metals and thus airborne metal fume contamination may have been deposited within the subject residence.

Sampling is most useful when conducted after the offsite research has been completed and knowledge of the nature, extent, and characteristics of the fire have been documented, reviewed, and the on-site visual and sensory inspection has been completed. With the above research and tasks completed, the investigator should have sufficient information to make an initial hypothesis as to the impact category from S&PC to the residence; the established categories are listed below:

1. Unaffected/Remediated
2. Mild
3. Moderate
4. Significant
5. Severe

The investigator may have a challenging time deciding between categories such as unaffected vs. mild, or moderate vs. significant. Usually, it is easier to differentiate between either end of the spectrum. This is where a well-planned sampling plan and methodology will help the investigator to obtain results that they can use to choose the best fitting category according to the preponderance of facts, observations, investigation evidence and analytical results.

The value of sample data is dependent on the ability of the investigator to collect (1) representative samples, (2) informative samples, (3) or intentionally biased samples to determine potential worst case. In the context of an analytical measurement, a good composite sample would be representative – meaning that the concentration of the analyte in the sample analyzed is a good representation of its levels in the material or system being tested. In a residential environment the goal of obtaining a *truly representative* sample is almost always impossible to accomplish, but with some supporting descriptions and estimation of semi-homogenous or similar areas, it may be somewhat representative (or at least, illustrative).

An example showing the conundrum and potential solution would be collecting a micro-vacuumed debris sample from carpet. Let us suppose that the carpet measured approximately 10 x 10 ft, or 100 ft^2 (or 9 m^2), and a 400 cm^2 (or 0.04 m$^{2)}$ sample was collected. Would a reputable investigator taking a random sample claim that the 0.04 m^2 area sampled or 0.44% of the carpet area sampled is truly representative? Is it is possible from a logical perspective to make such a claim? Is it economically feasible to collect sufficient samples to adopt a truly "representative" sampling goal?

The answer is to *collect informative samples*. In this case, one data point for our decision is the gross debris loading of the carpet in mass per 100 cm^2. Another relevant data point would be the percentage of the products of combustion in the micro-vacuumed sample. These are both important data points, especially if we know that the sample was semi-representative or purposely biased. Lastly, we combine the data obtained with the observations made: does the carpet look new or old, damaged, or undamaged, clean, or stained, have an odor or not.

The investigator may purposely select a biased sampling regime to collect samples from what appears to be, based on observation, the most impacted areas to see what the worst-case scenario may look like. For our carpet debris sample example, if we know that one area of the carpet was heavily loaded with debris before cleaning, and we revisit that area to sample after the carpet was cleaned it may allow us to answer this question: is the carpet now cleaned to an acceptable level?

If we have a hypothesis that states that if the most heavily loaded area of the carpet is clean (meets our assigned cleanliness criteria), then we may logically assume that the rest of the carpet is clean, strictly speaking the logic is good. The logical result stated may not stand up to the rigor necessary for a truly "scientific" or statistically accurate answer. This is why the investigation of an impacted residence is a combination of science and art. Any investigator who claims complete impartiality is not well informed or being honest.

Whatever the logic and method used, it must be well considered, and tailored to answer the question; is there contaminants of concern present, and if so, what are their extent and concentrations? The logical choice for the sample location selection, reliability of the collection method and procedure used must be documented in a narrative format and should include pictures and possibly videos.

There are currently few standards for combustion-by-products sample collections. However, investigators should use the most recent professional guides such as 2019 AIHA " Technical Guide for Wildfire Impact Assessments for the OEHS Professional" and 2012 ANSI/IESO/RIA-6001 "Evaluation of Heating, Ventilation and Air Conditioning (HVAC) Interior Surfaces to Determine the Presence of Fire-Related Particulates as a Result of a Fire in a Structure." It is the author's objective in creating this handbook that it will be one of the more informative sources for guidance on logically selecting and collecting appropriate samples from S&PC impacted residences.

Screening Levels

From our experience, when conducting micro-vacuuming sampling of carpet, levels of debris (mass) in carpets that are less than 10 mg/100 cm^2 are relatively clean. Those with 10 to 20 mg/100 cm^2 are moderately dirty and those with loads greater than 20 mg/100 cm^2 are heavy loaded. Therefore, if the mass of debris found is greater than 20 mg/100 cm^2 we would consider it contaminated (further analysis may be conducted to determine the percentage of products of combustion in the mass removed). Note that this is not an accepted national standard; this is an EIS established screening value. See Appendix D for Method EIS-CMV-001.

P
icture showing four 100 cm^2 areas that have been sampled by micro vacuuming using 25 mm cassette (shown).

Airborne VOCs[52]

VOC and particulate sample concentration may be compared to a percentage of the Permissible Exposure Limit (PEL) or Threshold Limit Value - Time Weighted Average (TLV-TWA). EIS has its own established guidelines for VOC concentration levels. It is our opinion that each identified chemical evaluated should be less than one one-hundredth (1/100) of any PEL or TLV concentration value for the chemical. If the measured VOC values found for each chemical are at least two orders of magnitude less than the most conservative value either the Permissible Exposure Limit or the Threshold Limit Value - Time Weighted Average for the substance found, we would generally consider the concentrations found to be acceptable. This is a generalization. Specific chemicals should always have a concentration below the EPA IRIS[53] criteria or residential risk-based screening criteria.

SUMMA Canister.

SUMMA Canister regulator (note it is under vacuum).

It is not reasonable to expect a non-detectable concentration of VOCs in today's residential environment when they are measured in parts per billion (ppbv). It is not uncommon to find concentrations of acetone, ethanol, isopropyl alcohol in 500 ppbv levels or more in normal homes.

Currently, there is no health standard for indoor total volatile organic compounds (TVOC) levels in non-industrial settings. One criterion for assessing TVOC uses the guideline of 500 $\mu g/m^3$ or 0.5 ppm set by the Leadership in Energy and Environmental Design (LEED) as a target level. We believe that a recommendation maximum of 3.0 ppm (3,000 ppbv) for total VOCs is reasonable, with a target of 1,000 ppbv or 1.0 ppm.

Most of the screening levels recommended were adapted from those published by the Contaminants of Potential Concern (COPC) Committee of the World Trade Center (WTC) Indoor Air Task Force Working Group. We use these screening levels because we prefer this reference scale for interpreting concentration results. The screening value concentration levels represent those levels above which there is increased probability of an adverse health impact resulting from exposure to the contaminants. The risk established for these screening levels was set by the COPC Committee Working Group at 1×10^{-4} (a 1 in ten thousand risk). We will present the reference screening levels for the different contaminants in the appropriate section. [54]

Another potentially useful set of screening levels was assembled by the California Department of Toxic Substances Control (DTSC) Human and Ecological Risk Office (HERO). Refer to *HERO HHRA Note Number: 3, DTSC-modified screening levels (DTSC-SLs)*, release date April 2019. This document includes screening levels for soils, tap water and ambient air. What it does not include is concentration levels for settled dust, which we will be using extensively.[55]

World Trade Center Contaminants of Potential Concern Health-Based Benchmarks

The following tables summarize the WTC-COPC health-based benchmarks. We have translated the values into different units to allow for comparison with analytical results. The units that we use are ppm, ppb, µg/g, ppt, µg/m³, pg/m³ mg/100 cm², µg/100 cm² and mg/kg.[56]

Table 4.1: The WTC-COPC Health-Based Benchmarks for Indoor Sources of Dioxin and PAH

Substance	Indoor Air
Dioxin	1 pg/m³
PAH	0.2 µg/m³

Table 4.2: The WTC-COPC Health-Based Benchmarks for Dioxin from Bulk Sources

	mg/kg or µg/g	ng/kg
2,3,7,8 TCDD *TEQ*	6.00×10^{-5}	60

Table 4.3: The WTC-COPC Health-Based Benchmarks for Metals from Bulk Sources

Substance	Screening Value mg/kg or µg/g
Inorganics	
Aluminum	75,000
Antimony	30
Arsenic	23
Barium	5,250
Beryllium	150
Cadmium	75
Chromium	225
Cobalt	1,500
Copper	3,000
Fluoride	4,500
Iron	22,500
Lead	400
Lithium	1,500
Manganese	1,500
Mercury	11
Molybdenum	375
Nickel	1,500
Nitrate	120,000
Selenium	375
Silver	375
Strontium	45,000
Thallium	5
Titanium	300,000
Uranium	225
Vanadium	525
Zinc	22,500

Table 4.4: The WTC-COPC Health-Based Benchmarks for Metals from Settled Dust Sources.

Substance	Wipe Samples Settled Dust Screening Values: Screening Value µg/m²	Screening Value µg/100 cm²
Aluminum	1,567,888	15,679
Antimony	627	6.3
Arsenic	387	3.9
Barium	109,752	1,098
Beryllium	3,136	31
Cadmium	1,557	16
Chromium	4,704	47
Cobalt	31,358	314
Copper	62,716	627
Iron	940,733	9,407
Lead	270	2.7
Manganese	31,358	314
Mercury	157	1.6
Nickel	31,358	314
Selenium	7,839	78
Silver	7,839	78
Thallium	110	1.1
Vanadium	10,975	110
Zinc	470,366	4,704

Table 4.5: Selected Sampling Methods – Advantages and Disadvantages [57]

Sampling Method	Advantages	Disadvantages	Comment
Tape Lifts	- Good collection efficiency for particles with typical dust loading - Defined sample area for analysis - Maintains particle integrity during collection - Little to no damage to captured particles	- Low collection efficiency for rough surfaces - Limited sampling area per tape lift	- Use clear Bio-Tape or sample media provided by the analyzing laboratory - Do not fold the tape
Wipe	- Able to sample relatively large area - Efficient for relatively smooth nonporous surfaces with low or heavy loading - Can be used for chemical analysis (e.g., PAHs)	- Not appropriate for porous surface - Liquid agents may degrade or solubilize the particles - Can remove finishes (e.g., paint), affecting analytical results	- Both dry and wet wipes can be used - Wet wiping is recommended due to higher collection efficiency
Microvac	- Easy sample preparation in the lab - Effective for collecting particles from porous and uneven surfaces - Permits the confirmatory identification of soot by TEM - Permits chemical analysis of collected material	- Ineffective for collecting particles from relative *smooth,* nonporous surfaces with low loading - Does not preserve positions of particles on the original surface - Can induce damage to delicate structures or agglomerates such as char and ash	- Collect at least visible amount of debris - 25 mm 0.45 um pore size MCE cassettes are recommended
Air	- Easy sample preparation in the lab - Can determine if combustion-by-products are airborne - Can have direct respiratory health implication - Can be used for soot confirmation by TEM	- Limited to airborne particles only - Contaminate concentrations will be significantly reduced if sampling is done after extensive period from the fire damage	- Use Air-O-Cell Sample Cassette - Collect at least 150 liters of air (e.g., 10 minutes sampling at 15 L/min)

Sampling Methods & Results

We begin by reviewing a few commonly used methods for the sampling of combustion-by-products, including Tape Lifts, Dry and Wet Wipes, Micro-Vacuuming, and Air Monitoring. Each method has its own advantages and disadvantages.

Tape Lift Sampling

Tape lift sampling, when used properly, gives semi-quantitative and qualitative data on PC, but it does not necessarily indicate the potential toxicity of the material. It is one tool to be used by experienced investigators. Used wisely it allows the investigator to identify areas of potential contamination. Tape lift results (combined with additional data concerning the spectrum of contaminants) may provide significant insights into the extent of residual impacts from S&PC.

Collecting Tape samples as part of a wildfire or structure fire investigation typically has two related goals. The first goal is to determine whether the particle types, concentrations, or ratio of combustion-generated residues indicate an *atypical impact above background*; secondly, if atypical, then to determine extent of contamination. These goals can be achieved by preserving the in-situ sample integrity of the deposited particles for direct light (optical) microscopic analysis without introducing significant sample alteration or particle loss.

Wildfire char is composed of large, irregular, mostly carbonaceous fragments of burned vegetation from leaves, twigs, and bark, and 95 percent of the burned particles associated with a wildfire range in size from approximately 2 µm to 5,000 µm (5 millimeters). The average size of the wildfire char, ash, and other particles found as infiltrated settled debris indoors ranges from approximately 2 µm to 50 µm.

A light microscope equipped with simultaneous reflected light (dark field) and transmitted or polarized light imaging capabilities covering a magnification range of 50x–500x can be used to identify vegetative char, ash, and other optically opaque particles, with detail of the particle's morphological properties.

Smoldering fires can generate copious amounts of organic compounds that condense as large aciniform soot clusters. When these compounds come into contact with cooler surfaces such as walls and ceiling, they produce chaining patterns and organic films.

The amounts of soluble and semi-volatile organic compounds, and inorganic chemical residues, in conjunction with recognizable surface depositional patterns, are critical to characterize the source of the combusted material when it is not obvious to interpret the extent of exposure. The presence or absence of specific indicator or signature particles characteristic of the type of fire can also help confirm the source or origin of the fire.

There are well established procedures for using a light microscope (equipped with reflected and transmitted light, bright field, dark field, and polarized light capability) to accurately analyze the optical properties of combustion particles that are characteristic of a wildfire or structure fire. [58]

EIS is not aware of any reputable authority or relevant literature which would support a proposition that tape lift be used as the sole means to determine the extent of contamination and remediation method.

Tape lift sampling is an effective sampling method for collecting particles from *relatively smooth non-porous surfaces with typical monolayer loading (desks, furniture, glass, and hard floors).*

For evaluation, soot is likely to be much more relevant than ash or char and is the only component that the EPA and others specifically consider to be hazardous. The TLV-TWA for carbon black (similar to soot) is 3 mg/m^3.

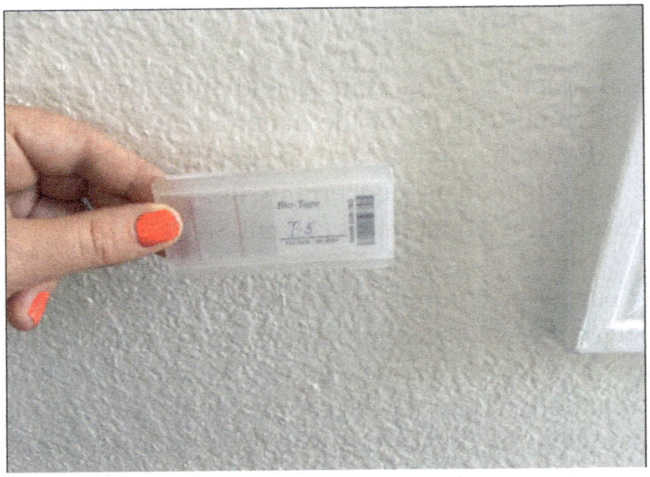

Tape lift sampling.

Regardless of the laboratory that the investigator uses, it is important to understand the evaluation method and limitations. [59,60]

Based on existing literature and a careful reading of the method; the unmodified ASTM 6602-13 Method (including the sample collection, sample preparation, and analysis procedures) is incompatible with (and was never designed to be used as) a method for the analysis of wildfire or structure fire combustion particles. This method was specifically optimized for the analysis of the residual "elemental" carbon end product residues formed at high controlled temperatures, while at the same time eliminating the unwanted "organic" carbon byproducts formed at lower temperatures. As a result, any attempt to apply this method specifically to the lower temperature "organic" carbon wildfire or structure fire combustion particles (especially soot condensate particles) is unpredictable and therefore inherently unreliable.[61]

As a comprehensive method example, the analysis provided by Environmental Analysis Associates (EAA), provides an initial qualitative dry examination of the color, texture, and presence or absence of large soot, char, or ash particles found on the entire tape lift slide preparation. The qualitative examination is followed by a quantitative estimate of the fire / combustion residue particles found in the sample. The adhesive tape lift sample analysis procedure preserves the integrity, size, and spatial distribution of the deposited dust, and is used to simultaneously estimate the area ratio percentage and the concentration of settled fire-related particles on sampled surfaces per unit area (cts/mm^2). [62]

Please note that EAA's in-house method and reporting format provides both qualitative and quantitative information but is not the only method that may be used for analysis. Other laboratories may have equivalent methods. At this time, we are not aware of any generally accepted methodology in the field of fire investigation.

EAA initial guidance information is based on data and observations collected over a 10-year period, and they provide an initial comparison chart dividing the quantitative data into four color-coded concentration ranges. The four ranges (*Typical-Low, Typical-Upper Background, Atypical, and Elevated)* are based on the total enumerated concentrations (area ratio %) and surface density (cts/mm^2) of fire-related particles and consider the background from other potential anthropogenic (man-made) sources. These ranges should be used for preliminary comparative estimates only. Determining if *Atypical* or *Elevated* measurements are related to a specific fire event must be based on all information gathered during the site investigation and should not be based on the laboratory data alone. The suggested color-coded concentration classification ranges are described below. [63,64]

The concentration levels do not automatically imply contamination, damage, a health impact, or the need for remedial action.[65] The ranges are also based on their multiplicative difference from database background or control samples (e.g., 3 or 10 times the background control range). It is suggested that wherever possible, onsite controls specific to the situation are used.

1. **Typical-low** - Normal background concentrations (approximately the 50th percentile range) found in non-impacted commercial and residential buildings.

2. **Typical–upper background** – Upper background range accounting for typical indoor or infiltration sources from cooking, candle burning, fireplaces, barbecues, roadway sources, etc. This upper range also accounts for laboratory analysis variability and the differential measurement values generated by differences in background dust loading.

3. **Atypical** – Concentrations measured approximately 3-10 times above the normal background range potentially associated with singular or multiple fire sources. It cannot be assumed that concentrations measured in this range are associated with a specific event without additional information regarding the site-specific conditions, the presence of indicator particles and the presence or absence of potential contributing sources.

4. **Elevated** – Concentrations measured approximately 10 times the normal background range or greater. Concentrations in this range are most likely associated with a specific fire event or chronic exposure to multiple events. Levels in this range still need to be evaluated in the context of the site-specific conditions, the presence or absence of specific indicator particles, and the presence or absence of potential contributing sources.

The background levels of fire / combustion particles in non-impacted commercial office and residential spaces are typically below 1% (area ratio%) with a range of 0.1% to 3%; with some locations as high as 5%.

Wildfire particles are primarily composed of char and ash particles and not aciniform soot. The lighter aciniform soot particles generated from the volatile plant resins are typical lofted high into the air in the crown of the fire and are not routinely found as condensed or settled debris inside structures unless the structure is within the burn zone.

Structure fires generate an entirely different distribution of soot, char, ash, and indicator particles from wildfire. The heated plume of a structure fire is usually contained within the building envelope, and the fuel-based and heated organic compounds form aciniform soot condensates that can deposit as larger clusters (e.g., 50-100μm), or even larger spider-web-like chains on cooler ceiling and wall surfaces within the structure. The depositional patterns generated by aciniform soot condensation are critical diagnostic indicators for an indoor fire and can be readily observed using an Optical Microscopy examination of tape lift samples collected directly from walls, ceilings, or contents.

Table 4.6: Total Area Ratio % and Numerical Surface Concentrations

Classification Range	Total Fire Residue (ratio%)	Total Fire Residue (cts/mm^2)
Atypical 3 - 10x background	3-10%	5-50
Typical - upper background	1-3%	1-5
Typical - low	<1%	<1

Wipe Sampling

Wipe sampling is an important technique for the estimation of contaminant deposition in buildings, homes, or outdoor surfaces as a source of possible human exposure. Numerous methods of wipe sampling exist, and each method has its own specification for the type of wipe material, wetting solvent, and analysis to be used, depending upon the contaminant of concern. [66] The objective of this handbook is to concisely summarize the recommended sampling techniques for a targeted list of agents that we believe are chemical markers for contamination created by smoke and products of combustion.[67]

Wipe sampling is one of the primary techniques for assessing surface contamination. When using the appropriate wipe sampling media coupled with an appropriate solvent (or used dry, as warranted), the type and amount of chemical on a particular surface can be identified by wiping a sufficient area of the surface and analyzing the wipe. This technique is a quick and easy means of determining what chemicals may reside on a surface. Most wipe sampling procedures are a manual process, and the pressure applied for each wiping procedure could vary significantly among field operators. Wipe sampling techniques are used for environmental sampling, industrial hygiene monitoring, assessment of remedial progress, compliance monitoring, and various other related applications.

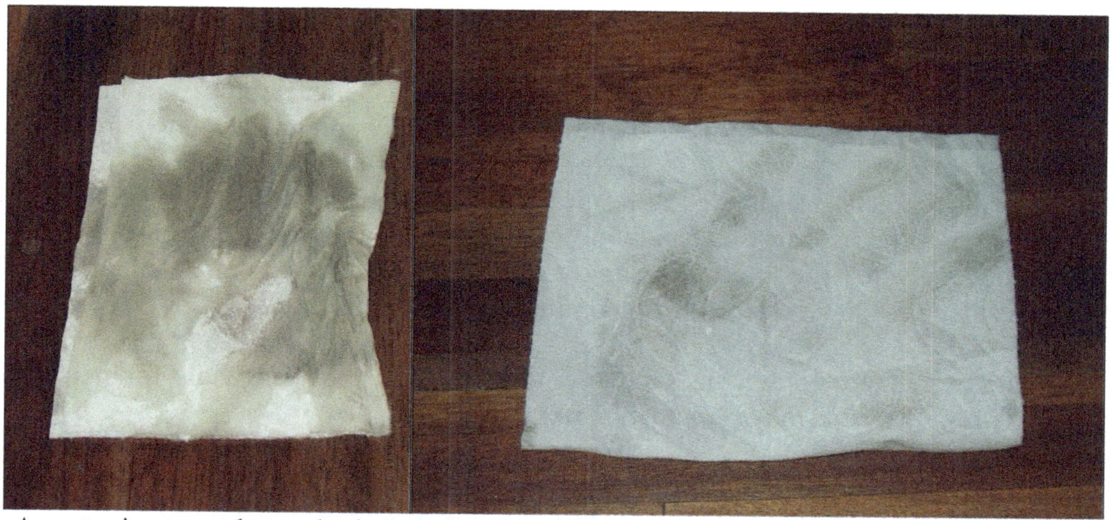

A wet wipe sample on the left, and a dry wipe sample on the right.

Procedures for the collection of contaminants from surfaces have some commonalities, including the use of wipe sampling media, wetting solvent, and the collection technique. However, wipe sampling procedures can vary widely, depending on the contaminant(s) of interest and the surface to be sampled.

Reliability of the sample results begins with accurate collection of a sample for analysis. Thus, the wipe sampling procedures used for a particular analyte on a given surface (including the proper combination of the wipe, sampling media, wetting agent, and collection procedure) are an integral aspect of whether or not the results generated will be representative of the contamination.

The American Society for Testing Materials (ASTM) has a lead-specific wipe sampling standard method for collecting settled dust on surfaces in and around buildings. A packaged, disposable towelette that is pre-moistened with a wetting solvent is used for the sample collection. Overlapping "S" and "Z" patterns are used when collecting the sample from an area of 100 cm^2. The area should be wiped vertically and then horizontally using firm strokes with minimal overlap.

The Industrial Hygiene Group at Brookhaven National Laboratory first developed the method now formalized as NIOSH Method 9100 (posted at www.cdc.gov/niosh/nmam) to determine lead and other metals in surface residue. A range of wipe materials can be used. Either 2" x 2" or 4" x 4" cotton gauze pads; ashless filter paper (1.5" to 4" in diameter); or pre-moistened wipes such as GhostWipes™ are appropriate for lead, beryllium, arsenic, cadmium, chromium, or nickel sampling.[68]

Approximately 1 to 2 milliliters (ml) of solvent such as DI water, isopropanol (IPA), ethanol, or n-hexane is used with the 100 cm^2 wipe. A 100 cm^2 area is supposed to be sampled by this method. The solvent used does not appear to be critical for the metal collection but can impact the sampling surface and should be chosen accordingly. Solvent selection for organic contaminants is critical; one must confer with the lab to get specific guidance on the type and source for the solvent (many times the lab will provide the appropriate solvent).

Table 4.7 summarizes the compound of interest, our recommended wipe media, solvent, and wipe area. The largest drawbacks we see from using wipe sampling in these investigations is that the relatively small area sampled, in relationship to the larger areas, characterizes the results for the larger area.

We recommend wiping an area of 400 cm^2 vs. area of 100 cm^2. We have found from experience that (1) sample media is very rarely overloaded, (2) sample solvent amounts of 4 ml vs. the usual 2 ml are sufficient for wetting a wipe of 400 cm^2, and (3) we can get a better signal from noise ratio for our sample by increasing the area wiped (thus we believe that our precision and accuracy are increased.)

After the analysis is complete, we divide the result by 4 to convert the results into areas of 100 cm^2 for standard comparisons. The user of this handbook may accept our advice to increase the wiped area or stick to the method as given. The important thing is to have the results expressed in units of 100 cm^2 so that comparisons can be made to accepted or recommended screening values.

Note: For containing and protecting the samples, it is OK to use plastic vials for non-organic wipe samples. *Always use glass vials for organic wipe samples.*

Table 4.7: Wipe Sampling Media, Solvent, and Suggested Area for Substances

Compound	Media	Solvent	Wipe area
Corrosivity /PH	Gauze Pad 3x3" or Alcohol Pad 4 x 3"	+96 % IPA or Ethanol	100 cm^2
PAH	Gauze Pads 3 x 3" or AA Wipes 9 x 9''	25% Acetone 75% Hexane	400 cm^2
Metals	Ghost Wipes 10 x 10 cm, or Gauze Pads 3 x 3"	DI Water or 50% IPA	400 cm^2
Soot/Dry	Kimwipes 30 x 30 cm or AA Wipes 9 x 9''	N/A	400 cm^2
Soot/Wet	AA Wipes 9 x 9'' (22 x 22 cm) or other Cellulose /Polyester Mix	91-96% IPA	400 cm^2

10 x 10 cm Template.

50 ml plastic vial for non-organic wipe samples.

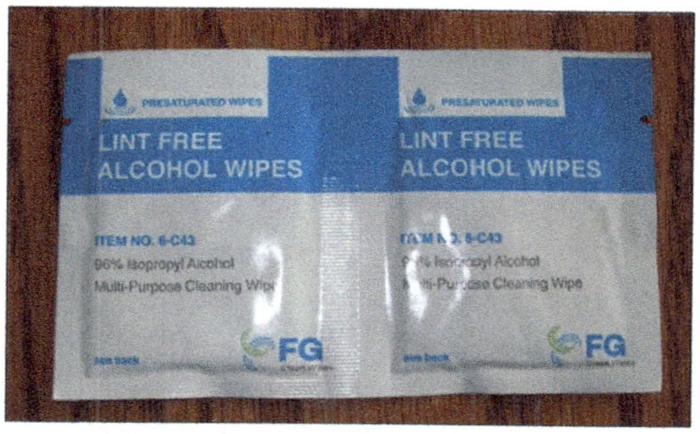

Alcohol Wipes.

Ghost Wipe for metals.

AA Wipe.

General Micro Vacuum Sampling

Micro-vacuumed samples can be collected from various locations to determine the amount of debris per unit area reported as mg of debris per 100 cm^2 area vacuumed. Samples should be collected by using either pre-weighed 37- or 25-mm sample cassettes and at least a ¼ hp vacuum pump with free flow rate of at least 20 lpm. Samples can be submitted to the lab for further analysis for specific contaminants of interest. Please confer with the laboratory for the sample filter media best suited to the analysis of interest.

Each cassette should be weighed either by the lab or investigator prior to and post micro-vacuuming to obtain the weight of the sample collected. We strongly recommend collecting a sample from an area of at least 400 cm^2 and then dividing the resultant weight by 4 to obtain the mass per 100 cm^2. This will reduce the sample precision error.

Carpet Micro Vacuum Sampling

Micro-vacuumed samples should be collected from carpeted locations to determine the amount of debris held in the carpet, reported as mg of debris per 100 cm^2 area vacuumed. Samples should be collected by using 25 mm sample cassettes and at least a ¼ hp vacuum pump with free flow rate of at least 20 lpm. Each cassette should be weighed prior to and post micro-vacuuming to obtain the sample weight collected. The scale used must be accurate to *at least* 1 mg (0.001 g). We strongly recommend collecting a sample from an area of at least 400 cm^2 and then dividing the resultant weight by 4 to obtain the mass per 100 cm^2. Refer to Appendix D for EIS-CMV-2022-001.

Sampling carpet.

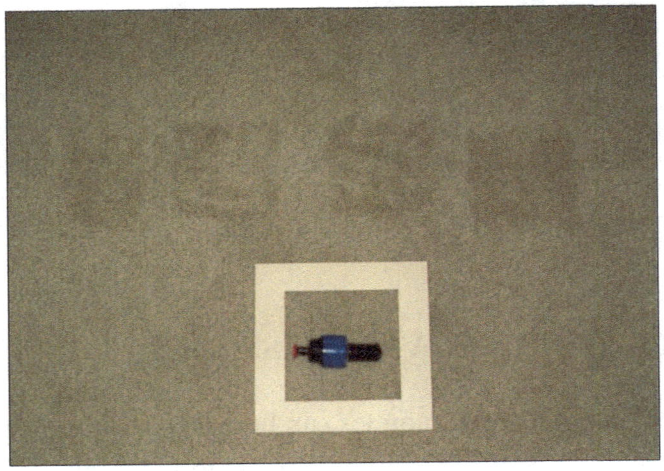

Micro-vacuum sample of carpet.

Picture showing four $100cm^2$ areas that have been sampled by micro vacuuming using 25mm cassette shown.

Carpet Micro-Vacuum Results Interpretation

From our experience, carpets with levels of debris (mass) less than ten mg per 100 cm^2 area (10 mg/100 cm^2) are relatively clean. Those with 10 to 20 mg/100 cm^2 are moderately dirty and those with loads greater than 20 mg/100 cm^2 are heavy loaded. Therefore, if the mass of debris found was greater than 20 mg/100 cm^2 we would consider it contaminated. If the loading is not believed to be from PC, further analysis may be conducted on the sample to determine the percentage of products of combustion in the mass collected.

Polycyclic Aromatic Hydrocarbons

Interpretation of PAH samples (collected using solvent, wipes, and an area of 100 cm^2 to 400 cm^2) can be made by reviewing your laboratory's analytical results. Always use the exact solvent recommended or provided by the analytical laboratory to avoid trace contaminants. If the laboratory found a concentration of total PAH to be equal to or exceeding 1.50 µg/100 cm^2 (WTC-COPC) screening level, we would consider that value significant and evidence of contamination.

Metals Determined by Wipe Sampling

Interpretation for samples collected using alcohol wipes and an area of 100 cm^2 to 400 cm^2 (we strongly recommend the larger sampling area if possible) can be determined by reviewing the laboratory analytical results. If the lab's results show a concentration of any of the metals in the following table exceeding the screening level, we would consider that value significant and evidence of contamination.

Table 4.8: Heavy Metals Subset - COPC

Metal Name	Screening Value µg/100 cm^2
Arsenic	3.9
Barium	1,097.5
Cadmium	15.6
Chromium	47.0
Lead	2.7
Mercury	1.6
Selenium	78.4
Silver	78.4

Bulk Sampling

Bulk samples may be collected for pH, asbestos, PAH, dioxin, and metals. Usually the mass needed for analysis (about 100 g for multiple analyte samples) precludes bulk sampling, but if sufficient material is available, it can be a cost-effective and more representative method to collect sample material. Measuring the pH of the debris is one especially useful analysis the investigator can conduct onsite using bulk material.

Bulk PC debris in sliding glass door track.

Bulk PC debris in windowsill.

Method to Determine pH

In a 16 x 150 mm (22 ml) test tube, add 5 g of PC debris and 5 ml of distilled water. Stir for 5 minutes with stainless steel spatula or cap and gently invert then revert test tube repeatedly during the time period. Let the suspension stand for about 5 minutes. Additional dilutions are allowed, if necessary, with hygroscopic samples or other problematic matrices; don't exceed 10 ml of DI water in total. After standing for 5 minutes, collect about a 0.5 ml of supernatant using disposable dropper or pipette; wet a pH test strip; read test strip and record results. The results will not be as acute as a sample sent to a lab, but it will give the investigator insight into whether the ash in the PC sample is likely to be corrosive.

Corrosivity Interpretation

Note – this interpretation is for samples collected using alcohol wipes and an area of 100 cm^2. A *highly* corrosive chemical has a pH of 2 or lower, or a pH of 12.5 or higher. A pH of 5 or less, or greater than 9, can be adverse for electrical equipment. We would therefore recommend that values with a pH of less than 5, or greater than 9 to be significant and require the appropriate response as determined by the actual pH if sensitive electrical equipment is contaminated.

Chloride Deposition and Electrical Equipment Damage

The following information was summarized from the publication, A Literature Review of the Effects of Smoke from a Fire on Electrical Equipment.[69]

Early reports on non-thermal damage to telecommunications were based on observations of the methods required to recover equipment from fire events. These reports categorize the level of damage to equipment based on the acid gas concentration on component surfaces, and *report that deposited mass of chloride can be a predictor of corrosive damage.* Reagor categorizes four levels of contamination[70] by mass of chloride deposited per unit area: clean at less than 2 $\mu g/cm^2$, easily cleaned and recovered below 30 $\mu g/cm^2$; difficult but possible to recover between 30 $\mu g/cm^2$ 90 $\mu g/cm^2$; and above 90 $\mu g/cm^2$ it becomes increasingly likely that the cost of cleaning the component will be greater than simply replacing it. Initially, Tewarson specified a minimum concentration of 0.98 $\mu g/cm^2$ coupled with relative humidity greater than 30 % at 20 °C (68 °F)[71].

We would consider chloride deposition on electrical equipment of greater than 200 ug/100 cm^2 to be significant.

Asbestos Sample Interpretation

The Permissible Exposure Limit (PEL) for asbestos is 0.1 fiber per cubic centimeter of air as an eight-hour time-weighted average (TWA), with an excursion limit (EL) of 1.0 asbestos fiber per cubic centimeter over a 30-minute period. If air samples exceed 0.01 f/cc clearance concentration level, we would consider that significant for asbestos.

OSHA does not, nor has it ever, recognized a 1% content in bulk materials as a safe amount of asbestos, but a content of 1% or greater, does; however, trigger specific, enhanced requirements in OSHA asbestos regulations. So, we consider a concentration of 1% asbestos in bulk material to be significant.

Dioxin Interpretation

The World Trade Center screening value for dioxin is 60 ng/kg TEQ. Florida adopted a residential soil Cleanup Target Level for dioxin (32 ng/kg) Toxic Equivalent Quantity. The draft USEPA document proposed a TEQ dioxin "Preliminary Remediation Goal" of 72 ng/kg for residential soil. So, there is a variety of potential interpretations of the results, but we will consider that value significant and evidence of contamination if the calculated TEQ was equal to or greater than 60 ng/kg.[72]

Guidelines for Sampling Based on Assumed Impact Category

We are listing the *minimum* specific sample types, numbers and location requirements based on the presumed observational category for most cases; there are exceptions to the rule that we will discuss later. The guidance is directed toward wildfire impacted residence, but the same method maybe used for other types of fires; instead of using entryways as the starting point for measuring sampling distances, use the geographic center of the presumed origin of the fire as the starting point. There is no restriction on the investigator collecting an additional number and type of samples according to their expertise, size of the residence, contextual information and their experience and knowledge.

Before one can conduct sampling, the investigator must make a general category decision because the number, method and extent of sampling will depend on the presumed impact category. Note that sampling requirements are cumulative, sequential and increase with impact severity. Therefore, the sample requirements for severe will normally incorporate those for unaffected- mild, mild-moderate, moderate-significant to significant-severe.

For Presumed Unaffected or Mild Impact:

1. Within a radius of 5 feet from each entryway, collect visual wipe and tape lifts from the midpoint of the wall (elevation approximately 4 feet high). Examine the dry or wet wipe sample for visual evidence of contamination. Make sure not to collect the tape lift sample from the area wiped.
2. Record visual results (suspected PC or None on wipe) and tape lift sample numbers; take photos of wipe & tape lift sample locations. Move out to a radius of 10 feet from the entryway and collect two more wipe and tape lift samples as in step 1.
3. Record visual results (suspected PC or None on wipe) and tape lift sample numbers; take photos of wipe & tape lift sample locations.
4. If the home has carpet on the main level, collect composite micro-vacuumed sample of the carpet for total debris; take photos of sample location(s). Analyze sample(s) for mass (total debris weight) and PC.

For Presumed Moderate Impact:

1. Collect all samples required for unaffected or mildly impacted category above in addition to those listed below.
2. Collect two visual wipe samples from at least two wall locations presumed to have had the greatest impact on each floor. Record visual results (suspected PC or None on wipe).

3. For every positive visual wipe sample, collect at least one tape lift sample; for each negative visual wipe sample, collect one. Label tape lift samples.
4. On each floor collect visual wipe samples from at least two horizontal surface locations suspected to have deposits of PC. These could be counter tops, tabletops, desktops, or similar surfaces. Record visual results (suspected PC or None on wipe).
5. For every positive visual wipe sample area, collect at least one tape lift sample.
6. If the investigator collected positive visual wipe samples from horizontal surfaces, then collect at least 3 samples from the different locations for corrosivity and heavy metals.
7. If the home has carpet on multiple floor levels, collect a composite micro-vacuumed sample of the carpet from each level for debris; take photos of sampled locations. Analyze sample for mass (total debris weight) and PC.
8. Use a PID (reading in ppb) to measure TVOC concentrations on each floor of the home.
9. Record sample collected, number and location; take photos of sample locations.

For Presumed Significant Impact:

1. Collect all samples required for unaffected, mildly, and moderately impacted category above in addition to those listed below, unless the level of observable contamination is sufficient to categorize the residence.
2. If the investigator has no historical information on potential PAH contamination from the fire, they may choose to collect one or more composite wipe sample (400 cm^2 from each area) or a composite sample of PC debris.
3. If the investigator has no historical information on potential dioxin contamination from the fire, they may choose to collect one composite wipe sample (600 cm^2 from 3 areas) or a composite sample of PC debris.
4. If samples are not needed to determine the category designation, they may be used to define the extent of significant impact. We would encourage this action, but it would be a decision for the more experienced investigator.
5. Record sample collected, number and location; take photos of sample locations.

For Presumed Severe Impact:

1. Collect all samples required for unaffected, mildly, moderately, and significantly impacted category above, unless the level of observable contamination is sufficient to categorize the residence.
2. If samples are not needed to determine category designation, they may be used to define the extent of significant impact. We would encourage this action, but it would be a decision for the more experienced investigator.
3. There are no additional sampling requirements since the distinction for severely vs. significantly is the intensity of the impact (the levels of contamination) not the types of chemical and physical contamination found. The main difference is that a severely impacted residence will also have physical damage that includes areas of the home that were actually burned. This will be obvious from inspection.

A severely impacted neighborhood with a number of residences burned to the ground & cars incinerated.

Impact Category's Physical & Chemical Characteristics

Listed below are the visual observations and contamination levels that define each impact category. The investigator should select the category based on the preponderance of observations and positive data points that fit the category. We recommend that the investigator be conservative; where ambiguity exists and there is a difficult choice between two categories for best fit, select the more conservative classification.

Unaffected/Remediated:

1. No physical damage to the building, lot, and/or structure from fire or S&PC.
2. No signs of staining or heat damage to the exterior or interior of the structure.
3. No smoke odor sensed within the residence.
4. Ash, char, and soot (PC) as identified using tape lift sampling are at background levels adopted from EAA as less than or equal to 3% area ratio percent of tape slide or surface density measured of less than 5 counts per square millimeter (cts/mm^2).
5. Total volatile organic compounds (TVOCs) are at background levels which we define as less than 3 ppm, and no single VOC is greater than 1/100th of the permissible exposure limit (PEL), or threshold limit value, time weighted average (TLV-TWA).
6. Total PAHs are less than screening value of 150 ug/m^2 or 15 ng/cm^2.

7. TEQ for dioxin is less than screening value of 2 ng/m^2 or 0.2 pg/cm^2.
8. Heavy metals are less than World Trade Center Chemicals of Concern Screening Levels.
9. Asbestos fibers are less than 0.01 f/cc.

Mild:

1. No structural damage from the fire to the interior or exterior of the building.
2. Interior and exterior show no signs of damage from heat or staining from soot.
3. Smoke odor may be present inside, but it does not create significant mucus membrane irritation (no signs of irritation after 30-minute exposure).
4. S&PC, if present, are visually deposited only at entryways and unsealed openings of the structure.
5. Ash, char, and soot (PC) on vertical surfaces (walls) or ceilings, as identified using tape lift sampling, may exceed background, but are localized within a 10-foot radius from entryways and has not affected more than 10% of interior surfaces of the residence as measured by floor square footage.
6. Total volatile organic compounds (TVOCs) may exceed background levels but do not exceed 3 ppm and no single VOC is greater than $1/100^{th}$ of the permissible exposure limit (PEL), or threshold limit value, time weighted average (TLV-TWA).
7. Total PAHs are less than screening value of 150 ug/m^2 or 15 ng/cm^2.
8. Heavy metals are less than World Trade Center Chemicals of Concern Screening Levels.
9. Asbestos fibers are less than 0.01f/cc.
10. TEQ for dioxin is less then screening value of 2 ng/m^2 or 0.2 pg/cm^2.

Moderate:

1. No structural damage from the fire to the interior or exterior of the structure.
2. Interior and exterior show no signs of damage from heat but may have some staining from soot.
3. Smoke odor is present inside and may create mucus membrane irritation within a 15-to-30-minute exposure.
4. S&PC are visually deposited at entryways and unsealed openings of the structure in addition to other locations.
5. Ash, char, and soot (PC) on vertical surfaces (walls) or ceilings, as identified using tape lift sampling, may exceed background, and may have impacted 10 percent of the residence but less than 20 percent as measured by floor square footage.
6. Total volatile organic compounds (TVOCs) may exceed background levels but do not exceed 6 ppm. and no single VOC is greater than $1/100^{th}$ of the permissible exposure limit (PEL), or threshold limit value, time weighted average (TLV-TWA).
7. Total PAHs are less than screening value of 150 ug/m^2 or 15 ng/cm^2.
8. Heavy metals are less than World Trade Center Chemicals of Concern Screening Levels.
9. Asbestos fibers are less than 0.01 f/cc.
10. TEQ for dioxin is less then screening value of 2 ng/m^2 or 0.2 pg/cm^2.

Significant:

1. Fire may have damaged nonstructural portions of the building and surrounding areas such as attached decks or perimeter fences.
2. Interior or exterior may show signs of heat damage such as blistering, or stained paint, or warping of windows.
3. Smoke odor is heavy and may be irritating with exposures of less than 15 minutes.
4. S&PC are visually deposited at entryways and unsealed openings of the building.
5. Ash, char, and soot (PC) on vertical surfaces (walls) or ceilings, as identified using tape lift sampling, exceed background and has affected more than 20 percent of the residence as measured by floor square footage. S&PC are greater than 10% area ratio percent of tape slide or surface density measured is more than 20 counts per square millimeter (cts/mm^2).
6. Total volatile organic compounds (TVOCs) may exceed background levels and individual VOC compounds may be greater than 1/100th of the permissible exposure limit (PEL), or threshold limit value, time weighted average (TLV-TWA).
7. Total PAHs may exceed the screening value of 150 ug/m^2 or 15 ng/cm^2.
8. Heavy metals concentrations may exceed World Trade Center Chemicals of Concern Screening Levels.
9. Asbestos fibers may exceed 0.01 f/cc.
10. TEQ for dioxin concentration may exceed the screening value of 2 ng/m^2 or 0.2 pg/cm^2.

Severe:

1. Fire will have damaged structural or nonstructural portions of the building and surrounding areas such as attached decks or perimeter fences.
2. Interior or exterior may show signs of heat damage such as blistering, or stained paint, or warping of windows.
3. Smoke odor is heavy and may be irritating with exposures of less than 5 minutes.
4. S&PC are visually deposited at entryways and unsealed openings of the building.
5. Ash, char, and soot (PC) on vertical surfaces (walls) or ceilings, as identified using tape lift sampling, exceed background and has affected more than 20 percent of the residence as measured by square footage. S&PC are greater than 10% area ratio percent of tape slide or surface density measured is more than 20 counts per square millimeter (cts/mm^2).
6. Total volatile organic compounds (TVOCs) may exceed background levels and individual VOC compounds may be greater than 1/100th of the permissible exposure limit (PEL), or threshold limit value, time weighted average (TLV-TWA).
7. Total PAHs may exceed the screening value of 150 ug/m^2 or 15 ng/cm^2.
8. Heavy metals concentrations may exceed World Trade Center Chemicals of Concern Screening Levels.
9. Asbestos fibers may exceed 0.01 f/cc.
10. TEQ for dioxin concentration may exceed the screening value of 2 ng/m^2 or 0.2 pg/cm^2.

Deck 4 x 4 and portion of retaining wall burned.

Residence's neighbor homes completely destroyed.

An incinerated ATV in residence's yard.

When Sampling Is Not Necessary

We will now address a situation where sampling is not really necessary to support a conclusion that identifies an *impact category,* but sampling may be valuable to define the extent of contamination. You may by analogy use it to identify others that fit the situation where the source of the contamination is known, and the observational information is extremely clear.

Please review the following pictures, taken in October 2021 by John Schenne used with permission.

This photo shows the burned iPad on the wooden kitchen table. *Photo by John Schenne.*

This photo shows the actual damage to the table. *Photo by John Schenne.*

Soot covering the entire inside of residence. *Photo by John Schenne.*

Soot deposited on ceiling/residue after first cleaning pass. *Photo by John Schenne.*

Register loaded with oily soot. *Photo by John Schenne.*

These pictures are taken from a residence that had an iPad spontaneously combust. It was in a plastic vinyl case located on a wooden kitchen table. The resulting smoldering fire generated a significant amount of smoke that deposited an oily soot on walls and ceilings. The kitchen was located on the ground floor level of the home; two bedrooms and a bathroom were located on the second floor. The home had a basement in which approximately 50% was finished. Please note that the author was called in to assist after significant but unsuccessful remediation activities had been conducted.

Let's examine the observations evident that we had to determine the category that this home should have been placed.

The residence had widespread soot on walls, ceilings, and in the gas forced air HVAC system. The HVAC ducts were coated with an oily soot. Immediately after the fire there was a strong odor of smoke, and the interior atmosphere was irritating to mucus membranes. We know from literature research that smoldering fires, especially from electrical equipment generate toxic smoke. A scientific article, *Toxic fluoride gas emissions from lithium-ion battery fires*, described emissions from a similar fire, and noted in the abstract:

> *Lithium-ion battery fires generate intense heat and considerable amounts of gas and smoke. Although the emission of toxic gases can be a larger threat than the heat, the knowledge of such emissions is limited. The results have been validated using two independent measurement techniques and show that large amounts of hydrogen fluoride (HF) may be generated, ranging between 20 and 200 mg/Wh of nominal battery energy capacity. In addition, 15–22 mg/Wh of another potentially toxic gas, phosphoryl fluoride (POF3), was measured in some of the fire tests*[73]

We already know from research that plastic fires have the capability to generate VOCs, PAH, dioxins, and other toxic and irritating compounds. These compounds contribute to the toxic/irritant quality of smoke gas atmospheres.

We would argue that the residence was significantly impacted from the fire (by observational information; knowing the source of the fire; and our research on the products of combustion from those sources). Without collecting a single sample, we would categorize this impact as significant, furthermore we would *also say that there are some characteristics of severe impact* since the residence sustained actual damage to components from heat or the fire. The residence was not habitable until remediation & restoration. This residence will not only require professional cleaning / remediation, but also some demolition and reconstruction for restoration to be completed.

To properly remediate this home, it will be necessary to remove drywall and insulation within an approximate radius of 15 feet from the fire source and replace the furnace and associated ducting. Drywall that was not replaced was first cleaned, sealed, and then painted.

Stud Bays & Complex Sampling

Arguments have been made to sample stud bays, the wall cavities usually about 14.5 inches wide between studs 2 x 4" or 2 x 6" wall cavities, for S&PC such as soot, char, and ash. We believe that there is little benefit to conduct such sampling except in very specific cases listed below.

1) No class I vapor retarder was installed in exterior walls, and fire had extreme winds associated with it that may have pressurized the residence and driven S&PC into stud bays.
2) The home was severely impacted and partially burned. S&PC may have migrated into the remaining part of the structure through the stud bays.

In situation (1) where there is no class I vapor retarder in the exterior walls, there exists the possibility that enough S&PC was driven into the walls to have contaminated the insulation in the cavity and that this contamination will result in a contaminate reservoir. The potential for the reservoir to continue off-gassing odors and potentially irritating vapors into the home must be addressed. If there is no vapor retarder, then the reservoir of contamination may pose a significant problem and necessitate the demolition of the dry wall to remove contaminated insulation and install a Class I barrier. Note that the contamination that we are most concerned with is soot and char, not ash, which usually does not contain significant amounts of adsorbed organic compounds that can be desorbed.

A visual inspection by either cutting an inspection hole in the wall or checking the wall at penetrations, such as electrical receptacles, will allow the investigation to determine if a vapor retarder exists. For more information about vapor retarders please see the next section.

Liquid water can accumulate inside exterior walls and in roof and crawl spaces. If enough water is present, rot and decay can cause significant damage. In areas where the climate is cold in the winter, the vapor retarder should be installed on the inner side of the wall near the warm interior space — or on the warm side in winter. In humid climates, or areas where there is extensive use of air-conditioning, if a vapor retarder is required, it should be installed on the exterior side of the wall. Building codes will specify when and where vapor retarders are installed.

A vapor retarder, sometimes called a vapor barrier, is normally added after insulation is installed to exterior walls in colder climates; this will be reversed in warm climates. The usual material for as vapor barrier is 6 mil polyethylene. A vapor retarder helps prevent water vapor from moving into building assemblies, like walls, where it can condense into liquid water within the structure.

A vapor retarder is required for structures conforming to the 2009 and newer editions of the International Code Council (ICC) building codes, summarized:

- The International Residential Code (IRC) defines vapor retarders as Class I, II or III based on how permeable they are to water vapor. The lower the permeability, the less water vapor that will pass through the vapor retarder.
  - **Class I – Very low permeability vapor retarders** – rated at 0.1 perms or less. Sheet polyethylene (e.g., Visqueen) or unperforated aluminum foil (FSK) are Class I vapor retarders.
  - **Class II – Low permeability vapor retarders** – rated greater than 0.1 perms and less than or equal to 1.0 perms. The kraft facing on batts qualify as a Class II vapor retarder.
  - **Class III – Medium permeability vapor retarders** – rated greater than 1.0 perms and less than or equal to 10 perms. Latex or enamel paint qualify as Class III vapor retarders.

In situation number (2), where there was an active fire in the structure and it was severely impacted, S&PC may have migrated into portions of the structure through the stud bays. Identifying if PC are in stud bays may be useful to help determine the extent of contamination, but the unfinished paper side of the drywall is not an easily sampled material.

The inspector should have a very good reason for conducting destructive sampling such as cutting holes in walls, but sometimes it is necessary. Be very careful about using tape lift sampling for stud bays, it is very difficult to collect a good sample on the surface of the drywall paper that is rough and grayish on the side that meets framing studs. To collect a sample of the rough grayish side for PC without getting the paper on the tape sample, it may be necessary to cut out a piece of the drywall and send it to the lab instead of conducting the tape lift.

It is important for the investigation to cut as many holes in walls as necessary to identify actual charring of the studs or other structural members, but this type of survey may be coordinated with the professional engineer who is tasked with determining the required demolition and reconstruction.

Hole in drywall to inspect floor joist (note charring).

Stud bay without Class I Vapor Retarding Barrier.

Bulk Samples of PC

With some wildfires there is so much PC debris deposited that the investigator may collect a bulk sample for analysis. Conducting analyses on bulk PC material may be a better choice than collecting wipe samples. Bulk samples may simply be collected using a new brush and dustpan: collect material from a number of locations in the residence to create a composite sample. Bulk material sampling will allow the investigator to collect a composite sample that is more homogeneous and may be more cost effective than collecting multiple wipe samples. One good bulk sample may be analyzed for metals, pH, PAH, and SVOCs more accurately and economically; and would be especially useful if the investigator will be assessing multiple residences impacted by the sample fire. A caveat is that the sample size will need to be fairly large, about 100 grams.

Smoke & Products of Combustion and Biological Contamination Sampling

If a residence has been impacted by S&PC and also experienced a water intrusion event (from the effort to extinguish the fire, or because pipes froze and burst, or from any other water source), it will increase the complexity of the investigation and recommendations for sampling and remediation. Let us look at some factors that affect the severity and size of the impact. There are three types of water: clean tap water or rainwater, brown water from sinks and showers, and black water from sewage. Refer to IICRC and their definitions of Categories 1, 2, and 3 water for more detail.[74]

If clean water is removed from the residence within 24 hours, and affected materials are quickly dried out, normally there will be little impact from mold growth; but with brown and black water, the problem is compounded, not only is mold an issue, but bacteria is too. Obviously all three types of water will cause damage to the residence structure and furnishings. Visual inspection and the use of a moisture meter will assist the investigator in determining the potential extent of the water intrusion event. The investigator may choose from a number of sampling methods such as spore trapping, culture, and PCR. Sampling is not usually required to identify the problem. There are usually visual and olfactory cues to note. The real value of sampling will be in post remediation or clearance, where aggressive air monitoring will be able to detect mold spores and bacterial contamination above background levels. An in-depth discussion of the investigation and sampling for biological contamination is beyond the scope of this book. Refer to IICRC, AIHA and ACGIH guidance for more information.[75]

SECTION 5.0 – REMEDIATION & RESTORATION RECOMMENDATIONS

Investigator's Responsibilities for Providing Recommendations

The creation of a remediation plan for the residence is not usually within the scope and responsibilities of the investigator providing the impact assessment, yet a general outline for the remediation has to be included in the investigators scope. Why? EIS has found by bitter experience that no other party has the experience, knowledge and expertise, objectivity and standing to provide the guidance needed to *establish the minimum necessary actions* to properly specify the remediation scope for S&PC contamination that may impact the occupant's health and comfort. It is our fervent hope that the guidance we are providing, used in conjunction with the *Guidelines for Fire & Smoke Damage Repair* and the methods we used to arrive at a defensible conclusion for minimum remediation & restoration requirements will obviate the historical lack of funding and appropriate scope for a successful remediation. For significantly impacted residences, we believe that the minimum remediation requirements cannot be set by the insurance company's adjuster or remediation - restoration contractor without input from a qualified, independent professional. The use of an independent/third-party inspection, and oversight by a qualified individual/firm employed by, or working on behalf of, the property owner, will help to ensure a satisfactory result.

Therefore, as professionals it is incumbent on the investigator to take on some of the responsibility for planning the remediation. We completely reject the notion that it is someone else's problem to determine the *general* remediation requirements. We feel that it is necessary and moral to translate our findings into tangible actions to take for a successful remediation of the residence. With that said, our recommendations are subservient to a professional engineer's evaluation of impacts and any necessary demolition and replacement of building components that they deem necessary. In fact, we would require the evaluation of *severely impacted residences* by a professional engineer.

Without such guidance the insurance company and remediation/remodeling contractor may not include necessary remediation and remodeling tasks. We must provide a check on the restoration-remediation contractor since most providing services for larger events *will not be local*, and therefore the residence owner, once they have paid for the remediation contactor's services, will find it impossible to take them to task for an inadequate job. Neither the insurance company nor the contractor or property owner is in a position to outline the *minimum* remediation requirements that will ensure the remediation job is competently completed.

To assist the investigator to complete this very difficult task, this handbook provides tools such as the Remediation & Restoration Protocol Matrix (available in Appendix E) and written advice for its use by the investigator.

Remediation and Restoration

The initial actions necessary for all S&PC-impacted residences is to remove gross S&PC debris and air out or scrub the atmosphere in the residence to remove contaminants. This may not be possible for significantly and severely impacted residences or in areas which have very high outdoor PC levels. These steps should precede all others to remove the source of further contamination of the residence and protect remediation contractor employees. This can be accomplished by HEPA vacuuming S&PC debris, opening the residence for natural ventilation, or providing air filters / scrubbers; further actions will depend on the impact category. It should be possible to clean mildly affected residences without removing all furniture from the home, whereas more heavily impacted residences will require that furnishings be first removed from areas to be cleaned; and the furnishings should be cleaned separately before being returned to the residence. Onsite storage pods are a good solution.

Cleaning, Remediation or Removal of Fire Generated S&PC

There is no single approach to cleaning that fits all fire impact situations. To determine the extent of required S&PC cleaning and restoration, use the R&R Matrix provided for general guidance and progress with a stepwise approach from least aggressive to more aggressive methods. Mildly and moderately impacted residences may be cleaned by their owners or cleaning contractors that would not normally be considered remediation companies. In those instances, cleanup may be limited to basic surface cleaning only. In cases where the residence's impact category has been determined to be significantly or severely impacted, the recommendation is to use a trained remediation contractor.

Remediation Contractor Requirements

The contracted remediation & restoration company should be proficient in the restoration of fire and smoke related impact. Proficiency is obtained through specialized education, training, and field experience which includes, but is not limited to fire and smoke impact, decontamination, and cleaning method.

Contractors acquire restoration skills through experience, industry technical training programs, and certification programs. They must understand residential construction and common building materials. Education and training on removal techniques are available through internationally recognized certification and training bodies such as Institute of Inspection, Cleaning, and Restoration Certification (IICRC). The contractor should be appropriately licensed and insured as required.

The remediation contractor should have trained crews that are qualified and know how to use air purifying respirators, protective garments (Level C PPE) and comply with the OSHA and any applicable local jurisdictional health and safety regulations.

Contractors that perform normal or emergency services should have an enforceable written agreement to authorize work prior to starting the remediation. This agreement should be executed by the property owner, or their agent, to authorize either a limited scope of services (such as stabilization) or full mitigation. The agreement should spell out the responsible parties, the intent of the work, terms of payment and most importantly the guidance and methods that will be used to confirm that the work has been satisfactorily completed.

The investigator's inspection report should allow the contractor to understand the minimum required procedures required to address the contamination and remediation level of effort. This will allow the contractor to determine the resources to be used, procedures, and the potential schedule. Enough detail should be collected by the investigator to allow the contractor to calculate the cost of services with the understanding that hidden or undetected damage or contamination may be found during the services; there must be a method to price out these changes.

Use of Clearance or Check Sampling

Clearance or Post-remediation sampling is important to ensure that the residence is truly **clean** and has been returned to a pre-fire unaffected state. Check sampling is another type of sampling that is especially important, but seldom conducted, and is used to determine the effectiveness of the ongoing remediation. Check sampling should be employed if one suspects that the remediation contactor is not meeting expectations. Final clearance sampling should be conducted to *ensure that the remediation contractor is truly producing a quality job and meets the evaluation criteria for remediated residences.*

Preferred testing methods include tape lift, wipe, or micro-vacuuming samples (or a combination thereof) to check the quality of the remediation job *prior to final payment* made to the remediation contractor. Similar sampling methods are also recommended for clearance sampling, with the addition of aggressive air monitoring. These tests can be completed relatively quickly so an answer can be obtained before the contractor demobilizes. We define aggressive air monitoring as employing a fan with an output of 2,000-2,500 cfm to move air in a room in which air monitoring is being conducted for char, ash and soot, or mold spores. See Appendix C for detailed description of evaluation criteria.

How to Interpret the Remediation and Restoration Matrix

The following guidance is provided to help interpret and use the remediation and restoration matrix. The investigators should have conducted their assessments and decided on a specific category of impact for the residence. Refer to the Remediation and Restoration Matrix and review the row for location and materials then the column for the category – the intersection provides the recommended minimum action necessary.

The insurance company, remediation and or restoration company, homeowner or their representative should review the necessary actions for specific locations and materials. The *stated R&R actions established a minimum,* not maximum response. The actual remediation contractor may request additional actions but should substantiate their need by referencing the IICRC or RIA guidelines. The remediation contractor should be held accountable for conducting the minimum necessary actions listed and meeting the performance standard listed in the category for unaffected or remediated residences. Owners should insist that insurance companies pay for no less than the minimum suggested activities and can incorporate the guidelines elucidated in this handbook for reference. *Owners should not allow remediation contractors to try to clean or remediate items that have been listed for disposal* – it would be a waste of time, effort, and money.

General Remediation & Restoration for Mild to Moderately Impacted Residences

Below are general guidelines that are applicable to most mild to moderately impacted category restoration and remediation projects but may not necessarily be performed in the order presented. Changes are dictated based on each individual project. Where more detailed guidance is available such as those in the IICRC standards use the more detailed guidance.

1. Establish controlled access to work areas.
2. Ensure all workers entering the work area are aware of the hazards and are properly trained.
3. Use proper PPE, to include gloves and N95 or better Respirators, where necessary.

4. NO EATING OR DRINKING in the work area.
5. Ensure Safety Data Sheets (SDS) for all cleaning products used are available for review in the workplace.
6. Remove or protect all contents that cannot be cleaned in place.
7. Perform cleaning for combustion byproducts via HEPA vacuuming that may be followed by surface wiping as necessary.
8. The HEPA filter vacuum cleaner should be equipped with a filter system capable of collecting and retaining dusts.
9. Contractor should work with the owner to coordinate subcontractors for HVAC, duct cleaning, carpet cleaning and any other needed services so that the sequence of required tasks are completed in a fashion to limit cross contamination.
10. Isolation should be performed to prevent migration of combustion byproducts, dust, and debris into other areas of the property.
11. Perform check sampling as needed and post-restoration verification sampling to ensure the remediation is successful.

Scope of Work

Work Area Preparation

If possible, turn off HVAC system and use supplemental heating and cooling. This way the residence can have temperature control and ventilation during the process of removing contamination. Controls should be established to prevent the spread of PC by the system.

1. HVAC ventilation shall be shut down, and preventative measures shall be taken to prevent accidental start-ups. Intake and exhaust openings shall be sealed with at least one layer of 6-mil poly sheeting. The seals shall be installed in such a manner as to guarantee that the seals shall remain in place for the duration of the project.
2. Movable Objects - All contents and moveable objects should be removed from the work area unless they can be cleaned in-place. Use a HEPA vacuum to clean non-porous surfaces.

Combustion Byproduct Cleaning Protocol

Perform cleaning of impacted areas and surfaces in the work areas using the following work procedures and practices:

1. Use a HEPA vacuum to clean horizontal and vertical porous surfaces of impacted areas/materials.
2. HEPA-vacuum surfaces and items located at the top of the room and proceed downward.
3. Wipe clean all non-porous surfaces with an applicable cleaning detergent and all porous surfaces with a dry soot removal sponge.
4. After completion of restoration cleaning, have the contracted Certified Industrial Hygienist conduct a visual inspection with limited combustion byproduct surface tape lift sampling and aggressive air monitoring.

Activities for this project include the cleaning of the following specific areas/surfaces/materials that have been identified as potentially impacted by char, ash, and/or soot:

Interior

Entry points at windows and exterior doors:

HEPA vacuum exposed horizontal surfaces within 10 to 20 feet of entry points.

Following vacuuming, wipe clean exposed vertical and horizontal surfaces within 10 to 20 feet of entry points.

Wipe clean exposed vertical and ceiling surfaces within 10 to 20 feet of entry points.

Distances may be adjusted according to sampling results.

Exposed floors:

HEPA vacuum exposed floors.

Following vacuuming, clean carpets if determined to be cleanable.

Following vacuuming, wipe clean exposed resilient flooring (tile, laminate, sheet vinyl, etc.).

HVAC System:

The HVAC should be cleaned by an appropriately licensed HVAC contractor, and in accordance with National Air Duct Cleaners Association (NADCA) guidelines and standards. Replace HVAC return filters as often as necessary.

Attic:

Remove all loose insulation.

For bat insulation with Kraft surface HEPA vacuum surfaces.

Vacuum exposed horizontal surfaces throughout.

Replace removed insulation.

Garage

Remove or protect all contents that cannot be cleaned in place.

Vacuum exposed horizontal surfaces throughout.

Following vacuuming, wipe clean exposed horizontal and vertical surfaces throughout.

Exterior

Clean the exterior using low-pressure wash. Wash clean and rinse, working from top to bottom, including but not necessarily limited to roof components, walls, doors, windowsills, window troughs, etc.

Post-Restoration Verification Criteria

A post-restoration verification assessment and clearance sampling should be performed following completion of cleaning efforts. Only after all of the specified criteria are met should the restoration be considered complete and the areas clean. If any of the criteria are not met, then the restoration is not acceptable, and re-cleaning will be conducted at the contractor's expense to include all costs associated with re-sampling.

The criteria for post-restoration verification residences should meet the definition of unaffected residences. Clearance sampling should be conducted to ensure remediation is successful, including (1) no visible signs of fire damage; (2) no visible signs of smoke impacts, including no dust or particulates; and (3) all work has been completed in compliance with the restoration protocol. Post-restoration sampling confirms remediation has "passed" based on the evaluation criteria for remediated residences (see Appendix C.)

Further Reading

Institute of Inspection Cleaning and Restoration draft Guidelines

2nd Edition RIA Guidelines for Fire & Smoke Damage Repair

Appendices

Appendix A: Excerpts from Marshall Fire Operations After-Action Report

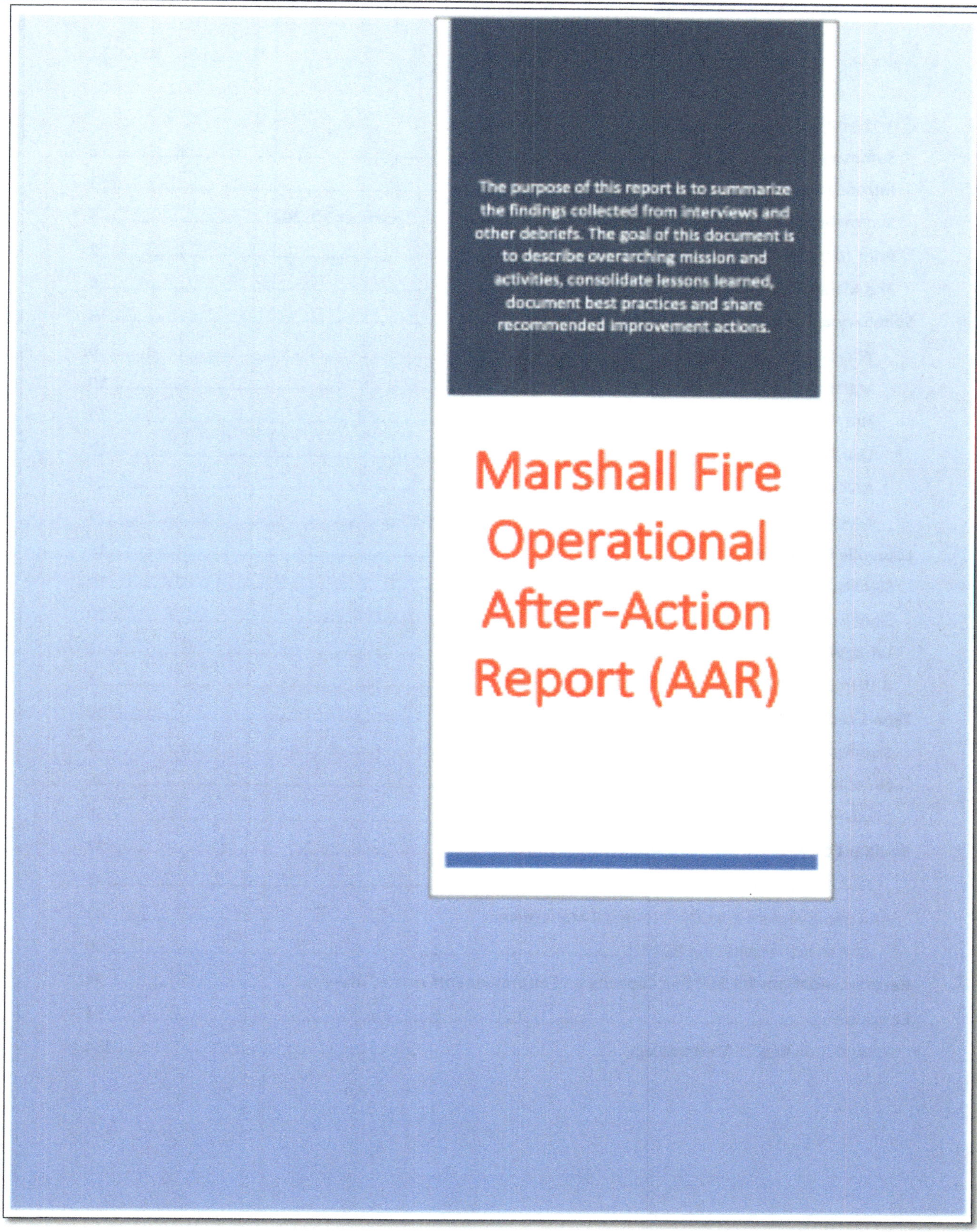

The purpose of this report is to summarize the findings collected from interviews and other debriefs. The goal of this document is to describe overarching mission and activities, consolidate lessons learned, document best practices and share recommended improvement actions.

Marshall Fire Operational After-Action Report (AAR)

Marshall Fire Operational AAR December 30, 2022

Contents

1

Summary

The After-Action Review (AAR) process is useful for reflecting upon an incident after its completion, fostering organizational learning, identifying areas of improvement, and learning from successes, challenges and failures. The ultimate goal of the AAR process is to improve response to future disasters.

Conducting the AAR process takes into consideration the complexity of the incident or disaster, and the duration and size of the operational response to design an effective process. The Marshall Fire AAR effort brought together decision makers and tactical resources into targeted AAR discussions to obtain critical elements of information, establish timelines, recount experiences based on areas of effort and extract what went well and what gaps or needs were present. The AAR process focuses on the operational response to the Marshall Fire for the first 12-36 hours.

The AAR begins with an introduction describing the weather conditions and the operational environment that existed at the time of the Marshall Fire. The additional sections provide a timeline of events when possible, highlighting a summary of significant events based upon the input of individuals involved, defining what went well, identifying what did not go well and integrating recommendations into an improvement plan. The improvement plan addresses challenges that were faced or actions that did not go well. The gaps identified can be addressed through planning efforts, applying organizational capacity, dedicating resources, providing equipment, staffing, training and developing exercises to test and evaluate effectiveness of improvements made.

The improvement plan design identifies the capability area, observations from participants about what did not go well, and corrective action recommendations, and it identifies the lead agency, the name of the responsible person to lead the corrective action and a time for completion. The lead agency and person are responsible for facilitating or leading the corrective action program or process. Most if not all corrective action programs and processes identified in this report require the involvement of multiple jurisdictions and the engagement of their leadership, operational supervisors and personnel to be successful. Implementing the improvement planning effort is the most important step in the AAR process for continuous improvement to be successful in addressing gaps.

The After-Action Review provides analysis for lessons learned, best practices and recommendations for future planning, training, and exercise development. As improvement actions are identified and addressed, it is important that any relevant plans, policies and procedures are updated accordingly.

2

Introduction

Summary of the weather conditions for the Marshall Fire on December 30, 2021.

https://www.weather.gov/bou/HighWinds12_30_2021

On Thursday, December 30, 2021, the devastating Marshall Fire roared through unincorporated Boulder County, Superior and Louisville, Colorado. The fire started south of Boulder, Colorado and was fanned by intense winds along the Front Range Foothills. Wind gusts from 70 to 100 mph occurred right at the base of the foothills, including in Boulder and along Highway 93 south toward Golden. The strong winds fanned a destructive grass fire which originated near Marshall in unincorporated Boulder County, and then quickly spread east to Superior and Louisville. At last count, the Marshall Fire destroyed 1,084 homes, and seven commercial structures and damaged 149 homes and 30 commercial structures.

The very strong winds developed in the mid-morning hours on Thursday as the result of a mountain wave that developed from very strong westerly winds racing over the Front Range Mountains and Foothills. The mountain wave remained nearly unchanged through the rest of the day, resulting in very persistent and extremely high winds that was focused very close to the base of the foothills, along Highway 93 and points east to around Superior and, Louisville. It takes just the right combination of meteorological parameters, including stability, wind shear, and wind magnitude, to create a powerful and damaging windstorm like this one.

The second component that contributed to this disaster was the lack of precipitation during the latter half of 2021. The Front Range experienced a very wet first half of the year with much greater than normal precipitation and lush and tall grass growth. However, starting around July, a persistently dry weather pattern set up and held firm through the entire fall and early winter. Grasses, while typically dry this time of year, were exceptionally dry as very little snow had occurred through the entire fall season. The temperature and precipitation from July 1 through December 29 ranks as the 2nd warmest and by far the driest in the Denver's recorded history (since 1872). This same time period for Boulder was ranked 2nd warmest for precipitation, while 13th driest in recorded history.

Prior to the Marshall Fire

At 10:15 a.m. the Lefthand Fire Protection District responded to the Middle Fork fire north of Boulder, and the weather conditions and high winds caused rapid growth of that fire. The Middle Fork fire attracted multiple fire departments from Allenspark, Hygiene, Lyons, Boulder Mountain Fire Protection District, and Boulder County Sheriff Fire Management Team to provide fire suppression and 15 Sheriff's Office Deputies to evacuate residents in the area. First responders reported fire spreading quickly with high winds as they started to engage in fighting the fire and initiating evacuations. The Boulder Incident Management Team (IMT3) was requested and responded to the scene. The 911 center was receiving hundreds of calls reporting the fire and launching emergency notifications through the Everbridge system to residents. An evacuation site was established in Longmont at the Memorial Building for evacuees of the fire. The Boulder Emergency Operations Center was not activated, but two staff members were supporting the IMT3 call out process and evacuation site efforts.

The City of Boulder Fire Department also was fighting wildfires during this time and into the afternoon along their western flank of the city.

The Marshall Fire Begins

At 11:05 a.m. the Boulder County Sheriff's 911 Center dispatched Mountain View Fire Protection District and Sheriff's Office Deputies to a line down near Marshall Road and Eldorado Springs Road. A fire had developed near that same area, and it spread rapidly in cured grass fuels (dry grass materials) with relative humidity around 20% and winds up to 100 mph. The fire, now known as the Marshall Fire, began consuming structures almost immediately. Resources were directed from the Middle Fork Fire at 11:15 a.m. due to the Marshall Fire starting. Fire crews discovered a shed on fire in the area and begin to attack the fire. The fire was spreading in other areas, but the spread initially was not detected because the wind was causing a tremendous amount of smoke, dirt, dust, and debris that concealed the fire's growth and direction. Once crews determined that the fire was spreading east, the fire attack broadened. Law enforcement began to call for evacuations immediately and started the door-to-door evacuation process. At 11:30 a.m. the Emergency Operations Center started to open, and notifications were made to Emergency Support Functions to activate for the Marshall Fire.

Evacuations were being done by deputies and fire firefighters in the area of Marshall Road, 76th Street, and Cherryvale Road. The 911 center saw a large increase in calls to report the fire and in calls for assistance. As the fire progressed towards the town of Superior and the large commercial area around Costco and Target, it was evident that the fire was going to significantly impact Superior and appeared likely to directly impact the city of Louisville. At 2:30 p.m., evacuation of Avista Hospital began. Evacuations continued, and coordination between the IMT3 and the Boulder County Law Branch was critical for successful evacuation of residents. As the evening continued, the devastation further extended into the town of Superior and the city of Louisville.

There were two fatalities during the Marshall Fire, one in the Marshall area and the other in the town of Superior. Both fatalities had been contacted by law enforcement during the evacuation process and advised to leave.

At approximately 6:30 p.m., the wind decreased to 25-30 mph and the fire behavior decreased in intensity and rate of spread. High winds continued into the evening, and by the morning of December 31 fire size was estimated at 6,000 acres.

The morning of December 31 brought a significant weather change: winds were less than 10 mph and snow was forecast for later in the day. The afternoon brought four to eight inches of snow which effectively stopped all forward spread of the wildland fire.

Summary of Events of the Operational Response

1024: Middle Fork starts and is drawing in fire resources and 15 deputies. Fire spreading quickly with high winds reported.

1105: Marshall Fire call is dispatched.

1123 Fire Command requests additional units to Marshall Fire. Fire moving east through property to homes.

1138: Evacuations start in Old Marshall and along Cherryvale north.

1150: Cherryvale Law Units begin evacuating South Vale and Wildflower Ranch.

1159: Fire begins running east down Marshall Rd. Middle Fork Incident Command calling for resources.

1200: BCSO fire command and law command tie in together and operational coordination established. Relocated ICP to 76th Street and Marshall Road.

1203: Fire is reported past water treatment plant moving at rapid rate toward Superior.

1206: Start evacuations for S 76th Sagamore, and Costco area. Flames reported at the back of Sagamore subdivision.

1215: Evacuation Orders issued through Everbridge system to unincorporated Boulder County and Superior and went to McCaslin. Door to door evacuations initiated for the next 30 minutes in the area.

1218: Costco evacuating, and fire is at Mohawk. Fire in parking lot of Costco and smoke obscuring everything.

1219: Flames have reached Sagamore.

1235: Mountain View Fire Department calls for Spanish Hills evacuations. Fire is there before evacuations.

1241: 911 Center started airing home addresses and was overwhelmed with reports.

1245: Fire reported behind Home Depot in Louisville.

1246: At 68th and South Boulder reports of fire in the area.

1250: Call for evacuations south of South Boulder Road also SCL Hospital evacuation, and McCaslin East to Sports Stable needs to evacuate.

1250: Decision is made to evacuate the entire town of Superior.

1256: Flatirons Mall contacted, and work started on getting access to use Nordstroms (closed/moved out) to take shelter from wind and set up ICP while waiting for Mobile Command Post.

1300: Law Division is established at Flatirons Mall. Teams are designated from arriving outside agencies for full Superior evacuation.

Teams of 10 are assigned radio designators of A-P and assigned to one of three tasks; Evacuations (overall evacuations of large areas), Critical or Tactical Evacuations (people needing help or in areas that had already been evacuated), and Road Closures/ Traffic control (including areas in Louisville).

1308 Louisville running operations and door to door evacuations.

1315: Mobile command post at the Flatirons Mall and the Law Branch was established boosting incident command from IMT3.

1402: Teams assigned to flush traffic eastbound only, at US36/ McCaslin and Rock Creek Pkwy/Coalton. Mountain View Fire assigned to help with evacuations at Urgent Care and then Avista.

1500: IMT provided planning, resource status, situation status, logistics, check in with divisions and branches, coordination with evacuation planning.

1520: Delegation of management of the fire to the Sheriff.

1530 Entire town of Superior Evacuated.

1537 Boulder Police SWAT Team assigned to assist Louisville with evacuations and began evacuating Zones 1, 2,3,4 and set their own staging at Monarch High.

1540: Delegated the fire to IMT3 as the Incident Commander.

1600: Delegated the fire to the state, evacuations and firefighting continuing.

1900: Wind shifting and making progress and the system started to catch up to the incident. South Metro Fire, North Metro Fire, Jefferson County Communication, State Emergency Operations Center and CO Division of Fire Prevention and Control all ordering resources.

2000 to 0700: The weather conditions are notably changed from the prior 9 hours. Fire crews continue to fight fires through the night. Evacuations continue door to door by law enforcement. Site access control planning begins to control access. BIMT continues to manage the incident and orders resources for the next day and completes incident action planning.

What Worked Well

Interagency Cooperation.

- Instructions and assignments were clear, and personnel knew exactly what was supposed to be done and were able to stay on the tasks at hand.
- Communications between multiple law agencies was effective because law enforcement had clear and concise information. Agencies used Law channel and cell phones for communication.
- Relationships with other agencies worked well. Pre-existing relationships, either from working on prior incidents together or through previous interactions, were great and key to operational successes.
- During extreme conditions and hardships not encountered on other incidents the commitment to work well together from all agencies was experienced throughout the operation.

Incident Command

- Establishing unified command by law enforcement with fire departments early on in both incidents increased effectiveness in making operational decisions and use of resources.
- Utility company presence and the development of a strong interface with the incident command post was effective in controlling utilities in the area, making it safe for first responders.
- IMT3 deployment and assumption of command added capacity to manage the complexity and escalation of the incident until the FEMA IMT1 team took the fire on January 1, 2022

Operational & Resource Management

- Assembling resources quickly and handing out assignments (from Longmont Police SWAT and Boulder Police SWAT) was critical to the effective door to door evacuations.
- Radio discipline within the law enforcement groups was excellent and prevented overloading operational channels preventing radio network failures. Personnel saved radio traffic for important information.
- There was great initiative. Numerous people saw a problem and took it upon themselves to rectify the situation as opposed to having to be ordered to. The ability to make decisions and resolve issues at the point of the problem created many operational successes and resolved significant challenges encountered throughout the incident.
- Diverting law and fire resources quickly from the Middle Fork Fire to the Marshall Fire assisted with bolstering the response to the Marshall Fire as resources were becoming scarce awaiting additional resources to arrive.
- The loss of life in any disaster is hard to accept, and tragically two people did lose their lives. Due to the speed of the fire spreading into multiple communities simultaneously the potential for many fatalities was present and fortunately did not manifest during this incident.
- Over 1,100 homes and businesses were lost in this fire, but thousands of homes were saved because of firefighting efforts.

Evacuations

- The initial alerts and warnings for the Middle Fork Fire and the Marshall Fire went out quickly and roadblocks were established immediately.
- Door to door evacuations needed to be executed immediately into the response and quickly throughout the Marshall Road 76th Street area. This included many residents and animals.
- Over 300 law enforcement resources assembled to perform door to door evacuations in Superior, Louisville, and unincorporated areas of Boulder County.
- The fire spread was so aggressive that in three locations it jumped Highway 36, which is a six-lane divided highway. The fire caused a traffic jam requiring all traffic to be re-routed by law enforcement.
- Traffic congestion was resolved at Coalton/Rock Creek by sending everyone east from Coalton in all lanes to Highway 36. This cleared out traffic jams in Rock Creek.
- Avista Hospital had to be evacuated in a short period of time due to the rapidly advancing fire. Through coordination of multiple agencies, evacuations were able to remove patients to other facilities. There was a phone call directly from an Avista Nurse (inquiring about evacuating the hospital). Having direct contact with employee from the facility provided real-time situational awareness for understanding the fire nearby. Due to information provided, the dispatcher advised to evacuate if in danger—same message provided to many callers.
- Multiple long-term care facilities also had to be evacuated at the same time. Residents were moved to other facilities due to effective facility emergency planning and transportation resources provided by ESF 1 Transportation.
- Additional dispatch staff brought into backfill, then identified a single dispatcher to manage evacuations and Everbridge.
- The Flatirons Crossing Mall provided a staging area large enough to support over 300 law enforcement personnel, 74 fire agencies and hundreds of fire personnel. This permitted an efficient incident base and operating space for the incident management teams operating the incident command post.
- The fire was moving faster than some of the evacuation orders could be developed and launched. Social media use by the community and messaging between the Town of Superior and City of Louisville assisted to provide messaging on social media about the fire.
- Communications to fire command was excellent between the 911 center to obtain information and relay critical updates.
- The evacuation of 37,500 people during a fast moving and multi-directional spreading fire in a relatively short period of time (3-4 hours) is unprecedented.

Communications/ 911 Center

The Sheriff's Communication Center provides dispatching services for all law enforcement agencies in Boulder County except the City of Boulder, University of Colorado, and the City of

Longmont. In addition, the center also provides 911 dispatching and alert and warning services for fire agencies and rescue agencies in Boulder County. The 911 services and dispatching services for the Middle Fork Fire and Marshall Fire were managed through this center. The Middle Fork Fire already had supervisors of the communications center starting to arrange for additional staff due to the size and hazardous nature of the fire.

The additional staffing was present or was already enroute, due to the dispatch all-page sent for assistance during the Middle Fork Fire. Having these additional dispatchers added capacity by adding multiple fire dispatchers and one supervisor working the incident facilitated splitting roles and responsibilities. Back-up staff to assist with Computer Aided Dispatch for Fire and Law channels were also made available.

There was strong room awareness by dispatchers, considering we were working in a temporary location with less than half the space and different layout and consoles available. Incoming additional dispatchers knew where to go and who to back up. Additional supervisors were also available to make decisions and assist with dispatching, making concise and appropriate instructions to callers, specific to each situation.

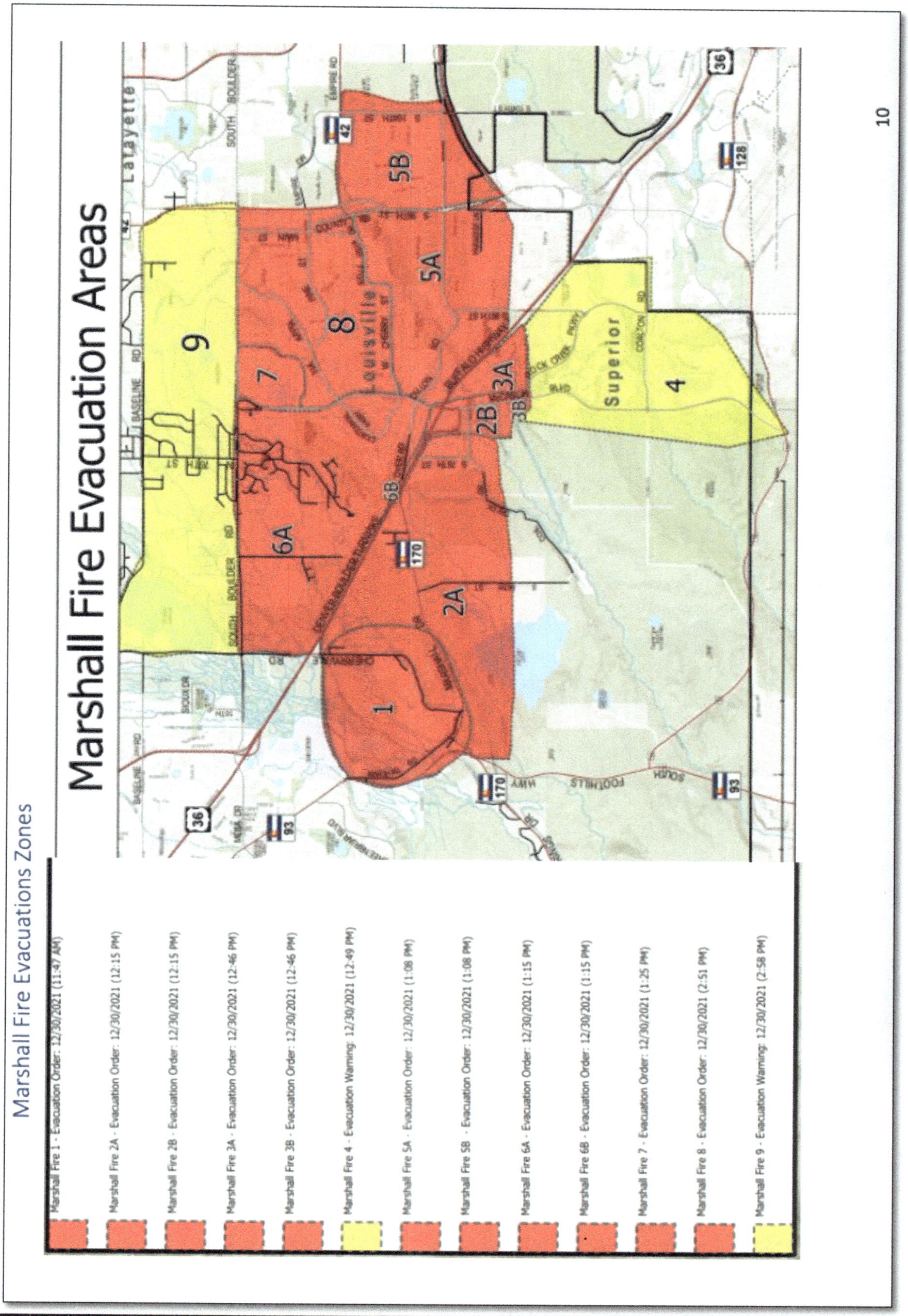
Marshall Fire Evacuations Zones
Marshall Fire Evacuation Areas
Louisville
Superior
Lafayette
1
2A
2B
3A
3B
4
5A
5B
6A
6B
7
8
9
Marshall Fire 1 - Evacuation Order: 12/30/2021 (11:47 AM)
Marshall Fire 2A - Evacuation Order: 12/30/2021 (12:15 PM)
Marshall Fire 2B - Evacuation Order: 12/30/2021 (12:15 PM)
Marshall Fire 3A - Evacuation Order: 12/30/2021 (12:46 PM)
Marshall Fire 3B - Evacuation Order: 12/30/2021 (12:46 PM)
Marshall Fire 4 - Evacuation Warning: 12/30/2021 (12:49 PM)
Marshall Fire 5A - Evacuation Order: 12/30/2021 (1:08 PM)
Marshall Fire 5B - Evacuation Order: 12/30/2021 (1:08 PM)
Marshall Fire 6A - Evacuation Order: 12/30/2021 (1:15 PM)
Marshall Fire 6B - Evacuation Order: 12/30/2021 (1:15 PM)
Marshall Fire 7 - Evacuation Order: 12/30/2021 (1:25 PM)
Marshall Fire 8 - Evacuation Order: 12/30/2021 (2:51 PM)
Marshall Fire 9 - Evacuation Warning: 12/30/2021 (2:58 PM)
10

Appendix B: Inspection Questionnaire & Checklist

Inspection Questionnaire and Checklist for Evaluation of Fire and Smoke Impacted Properties

4/15/2023

The following questionnaire and inspection checklist is to be used by the inspector to evaluate and record information and observations made prior to and during the onsite visit to the subject property. It should be used as an aid for conducting consistent evaluation of properties and is designed to be used before any remediation or mitigation activities begin. Answer all question with either; yes, no, not applicable (NA) or short comment.

I. Greater Area:

1. Name of Municipality or City:
2. Name of fire:
3. Date of fire:
4. Type of fire: Grass, Trees, Structural, Mixed, Other; Describe:
5. Wind direction during the fire:
6. Were maps of the area available/obtained showing extent of fire?
7. Was information available from local sources?
    a. Air monitoring during fire:
    b. Soil or debris sampling & analysis:
8. Unusual Facilities:
    a. Were any railcars, tanks or industrial buildings burned within 2 miles of subject property? (Y/N):
    b. Were any factories, warehouses or other facilities containing hazardous materials burned? (Y/N):
    c. If yes, list any potentially highly hazardous materials likely to be involved in fire because of such facilities:

II. Neighborhood of Subject Property:

1. Did the fire come within a mile or less of the subject property? (Y/N):
2. If Yes;
    a. Were homes burned? (Y/N):
    b. Were automobiles burned? (Y/N):
    c. If 'Yes', describe:

III. Adjacent to Subject Property:

1. Were homes burned within one block of the subject property? (Y/N):
2. Were automobiles burned within one block of the subject property? (Y/N):
3. Was the residence located downwind of the fire? (Y/N):

IV. Outside Perimeter of Subject Property:

1. Was the yard or fence surrounding the property burned? (Y/N):
2. If 'Yes', describe:

Record Sources of Offsite Information:

Onsite Evaluation of Residence

The following questionnaire-checklist is to be used by the inspector to collect and record observations and measurements made during the onsite inspection of the subject residence. This may be followed by sample collection.

Location / Address of Residence:
Date:
Inspector:
Owner or Agent:
    Name(s):
    Phone number(s):
    Email:
Interview the owner or agent:

1) Ask if they have any pictures/video pre, during or post event (Y/N):
2) Ask them to provide a summary of their experiences concerning the fire and impacts to their home:
3) Ask if their health & family members were impacted by the fire (Y/N):
    a. If 'Yes', describe:
4) Ask if any family member(s) are particularly sensitive to chemicals (Y/N):
    a. If 'Yes', describe:
5) Record important information acquired from the interview:

Exterior – Sketch: Make a sketch of the residence showing general exterior view showing cardinal points; identify cardinal direction of front and back, note egress locations. Take measurements to provide an estimate of the length and width of the residence. Take pictures or video of the front, back and sides and yard of the residence; note any visible fire, smoke or heat damage.

Add hand drawn sketch below or reference sketch file:

Initial Walkthrough Inspection & Observational Questions:

Garage:

Does the property have a garage? (Y/N):
If 'Yes', were PC noted in the garage? (Y/N):
Were PC threshold streaks visible below the garage door? (Y/N):
Is the garage physically connected to the structure? (Y/N):
Odor or irritation atmosphere in the garage atmosphere? (Y/N):
Take informative pictures/video.
Comments:

HVAC:

Describe the HVAC:
Inspect the system filter and describe its condition; note if filter has been replaced:
Inspect the interior of the HVAC and describe any visible debris, char, soot, or ash noted:
Take informative pictures/video.
Comments:

Attic:

Inspect the attic space:
Describe the type of insulation present:
Note any visible char, soot, or ash by vents or on insulation.
Take informative pictures/video.
Comments:

Wall Insulation:

Was the inspector able to see exposed wall insulation? (Y/N):
Was the inspector able to determine if a Class I vapor barriers is in place? (Y/N):
If so, does it appear to contain any fire related material? (Y/N):
Take informative pictures/video.
Comments:

Deck

Does the property have a deck? (Y/N):
If 'Yes', was any fire related damage noted? (Y/N):
Take informative pictures/video.
Comments:

Roof

Can the inspector view the roof? (Y/N):
If 'Yes', was any fire related damage noted? (Y/N):
Take informative pictures/video.
Comments:

Windows

What are the types of windows (vinyl, fiberglass wood)? :
Observable damage to windows or seals? (Y/N):
If 'Yes', describe damage:
Take informative pictures/video.
Comments:

Interior of Residence

Interior – Layout, Rooms and Materials: If possible, obtain an overhead layout view of each floor of the home. If that is not available create a rough sketch preferably on graph paper of each floor/level of the residence noting dimensions.

Record general observations by location; floors and rooms, take pictures/video:

For Each Floor Level Collect & Record the Following Information:

1) Number of rooms and identify each as kitchen (K), bathroom (B), bedroom (BR), living room (LR), dining room (DR), other______
2) For each room note the type of flooring; hard or carpeted, if carpeted record sq. ft. or % coverage:
3) For each room comment on any YES answer to the following & record description & information:
    a. Detectable smoke odor or irritation present? (Y/N):
    b. PC material or stains noted on floors, walls or windowsills? (Y/N):
    c. Was electrophoresis (ghosting) noted on any interior surface? (Y/N):
    d. Was thermophoresis noted on any interior surface? (Y/N):
    e. Were smoke webs or staining observed? (Y/N):

Appendix C: Evaluation Criteria for Remediated Homes

Evaluation Criteria for Remediated Residences

04/15/23

The following post remediation evaluation & acceptance criteria should be used by the investigator to evaluate residences that have been cleaned and remediated after being impacted by smoke and products of combustion. It is applicable for mildly to significantly impacted residences.

1. The remediation contractor should have followed Restorative Industry Association Guidelines for Fire and Smoke Damage Repair or other nationally recognized guidance.

   https://members.restorationindustry.org/store/viewproduct.aspx?id=4477272

   Contactor should be able to explain any deviations from the guidelines.

2. No vertical or horizontal surface, wall or ceiling or surface of furnishing should display any staining or dust related to smoke and products of combustion. This would be evident by observation or by dry wipe testing.

3. Char, Soot and Ash levels as defined in ASTM D6602-13 should be evaluated. The amounts found should be at background levels for non-impacted homes as measured by tape lift samples. For freshly remediated residences, use the low background level for common combustion-by-products of less than 1% area ratio with a surface density of less than 1 cts/mm^2. Contamination is considered likely if the reported concentration of combustion-by-products is significantly higher than the background.

4. If Char, Soot and Ash levels as defined in ASTM D6602-13 are evaluated by aggressive air monitoring using air-o-cell cassettes, the amounts found in the interior atmosphere should be at or below the low background level for common combustion-by-products of less than 1% area ratio with a surface density of less than 1 cts/mm^2.

5. If mold spores are evaluated by aggressive air monitoring using air-o-cell cassettes, the spores count found in the interior atmosphere should be at or below those collected from the outside atmosphere of the subject property. Therefore, contamination is considered likely if the reported concentration of mold spores in the residence is greater than the outside background levels.

6. Although not normally evaluated during post remediation check sampling, polycyclic-aromatic-hydrocarbon (PAH) concentration levels should be below the World Trade Center Chemicals of Concern screening level of 1.5ug/100cm^2. Appropriate areas in homes should be evaluated by wipe sampling of non-permeable surfaces.

7. Although not normally evaluated during post remediation check sampling, indoor air volatile organic chemical's (VOC) concentration levels for each identified chemical found must be less

than one percent (1.0%) of any listed PEL or TLV concentration. The recommended maximum for total volatile organic compounds (TVOC's) should be less than three parts per million (ppm).

8. Bulk dust should not be observable in living areas nor in any area in which material may become airborne and be reintroduced to interior home living spaces. This is especially important for ceilings, attic / loft locations and for the HVAC system.

9. Micro-vacuumed samples for carpet debris loading should not have a mass greater than 10 mg / 100 cm^2 and should not contain products of combustion greater than 3% of the relative mass.

10. Although not normally evaluated during post remediation check sampling, dust or chemical residues on hard surfaces should not have concentration levels of any World Trade Center Chemicals of Concern contaminants above established screening levels.

11. There should not be any odor of smoke or other non-natural odors sensed by occupants. This should extend for a period of 6 months from the remediation date completion.

12. The HVAC system should be checked and at a minimum have had the system filters replaced. For moderately and above impact categories it is recommended that forced air systems have been cleaned. Refer to **IESO/RIA Standard 6001, Evaluation of Heating, Ventilation and Air Conditioning (HVAC) Interior Surfaces to Determine the Presence of Fire-related Particulate as a Result of a Fire in a Structure, 2012 Edition.**

13. All permeable surfaces such as clothing, carpet, rugs, and furnishings should have been professionally cleaned or laundered and if damaged or not cleanable be disposed. Dispose of any articles that are damaged, likely to provide a contaminate reservoir or not reasonably cleanable.

Appendix D: EIS, Inc. Methods

Method Number: EIS-CMV-2022-001

Carpet Micro-Vacuum Sampling Technique

There are multiple techniques for collecting dust from surfaces using the micro-vacuuming technique. Our carpet sampling method combines some of the techniques presented by ASTM D5755-095 method for asbestos fiber collection with our experience collecting carpet dust samples. We believe that the method is best suited for extraction of dust from porous materials like carpet. This technique employs a vacuum to extracts dust from the full depth of the material not just the surface. The dust is captured into a sampling cassette, which may be submitted to a laboratory for further examination. We recommend that the cassettes be pre-weighted using a scale with the necessary sensitivity (0.001g). This allows for the determination of debris concentration in mg/100 cm^2 quickly.

We recommend collecting samples using 50 to 100 cm^2 surface templates (to normalize collection area) and to sample an area of 400 cm^2 (always record the actual area sampled). The sampling technique is relatively straightforward. It requires a sampling cassette (either 25 or 37 mm), a surface template, and a high flow sampling pump (vacuum pump) operating at approximately 20 L/min. A sampling kit can be purchased from SKC, Part Number: 225-9540. The kit includes 10 micro-vacuum cassettes, disposable 10 x 10 cm templates, Luer adapters, zip bags, and labels.

Select the carpeted area to be sampled and place a template on it to outline the sampling area. Take a sampling cassette and remove the front and back stopper; for a 37 mm cassette you will need to add a Luer adapter with a 1-inch-long piece of hose. Connect a sampling cassette to the vacuum pump and turn the pump on.

Start collecting the dust sample by moving the cassette tip against the surface. We suggest that the filter tip be moved in the following motions (see Figure 1):

1. A "North to South" zigzagging pattern, then
2. An "East to West" zigzagging pattern, then
3. A diagonal motion "Northwest to Southeast" zigzagging pattern, then finally
4. A "Northeast to Southwest" zigzagging pattern.

There should be four passes made on each 100 cm^2 area.

Before the vacuum pump is turned off, point the open end of the cassette to a vertical position, then turn of the vacuum pump and cap both ends the cassette. Once the sample is collected, record the sample ID, write down the surface area that was sampled and its location. Re-weigh the cassette to determine the loading. The cassette may be sent to a laboratory for further analysis for PC or other contaminants.

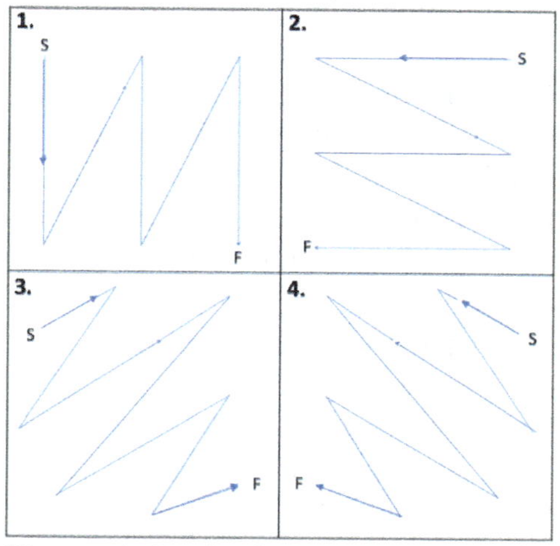

Figure 1. Paths for sample collection with starting locations (S) and finish locations (F) shown.

Method Number: EIS-AAM-2022-001

EIS Aggressive Air Monitoring Method for Mold Spores and Products of Combustion

Discussion:
This aggressive air monitoring sampling (AAM) method is to be used in residential room sizes from 64 to 400 ft^2 with nominal 8 to 10-ft-high ceilings. This approach requires the use of one 2,000 to 2,500 cfm square mixing fan and one vacuum pump sampler to be deployed in each room. It is specifically designed to be used with Zefon International air-o-cell cassettes. For larger rooms one may divide the space in half, or as many multiples as necessary as long as the maximum area of 400 ft^2 is not exceeded.

The mixing fan is employed during AAM to keep aerosolized particles airborne, so they are more likely to be collected by the air sampling filter cassette. The method requires one 20 to 22 in^2 box fan to be placed in the approximate center of the room, operated at high speed, and directed toward one of the walls of the room being sampled to create a circulating air pattern. The fan should be placed at an approximate distance of 2 ft perpendicularly from the stand holding the air-o-cell cassette. The stand should raise the cassette sampled to a height of approximately 3.5 ft. All distances should have a tolerance of +/- 10 percent.

The sample vacuum pump must be at least a ¼ hp vacuum pump, capable of producing a free flow rate of at least 20 lpm.

Method:
Select the room to be sampled; measure the length and width. If the room is larger than 400 ft^2 divide it in half or more to make spaces no larger than 400 ft^2. If the room is smaller than 64 ft^2 or has a ceiling higher than 10 ft, *do not use this method.*

Find the center of the room. Place the box fan in the center of the room. Point the fan at one of the walls. Place the sample stand 2 feet from the box fan and attach the sample cassette so that it is at a height of 3.5 feet above the floor. If furnishing or other encumbrances preclude the exact layout called for, try to place the fan and sample cassette as close to the assigned positions as possible. If necessary, divide the room into smaller spaces and collect multiple samples.

Record a simple sketch of the actual room size, location of the fan and sample cassette.

Use a rotometer with a valve to adjust the suction flow rate to be 15 lpm. When calibrating the flow, have a dedicated calibration air-o-cell cassette in place. Using a flexible hose connect the sample's vacuum pump to the air-o-cell cassette. Make sure that the protective tape is removed from the inlet and outlet of the cassette. Turn on the vacuum pump and time the run to be 10 minutes. Other times may be used, but make sure to note the actual time used. Please refer to recommendations from the air-o-cell cassette manufacturer.

Once the preselected time period is reached, turn off the sample pump, remove the sample cassette, record the sample number, location, flow rate, air volume and time on the laboratory's chain of custody form and then ship the sample to the selected laboratory for analysis.

Appendix E: Remediation & Restoration Protocol Matrix

Remediation & Restoration Protocol Matrix

4/15/23

#		Material	Category: Mild	Moderate	Significant	Severe
1	Attic	Insulation	Remove loose insulation & Replace; HEPA vacuum bat insulation if isolated contamination; Deodorize	Remove insulation; HEPA Vacuum clean; Deodorize & Replace	Remove insulation; HEPA Vacuum clean; Deodorize & Replace	Remove insulation; HEPA Vacuum clean; Encapsulate if necessary & Deodorize
2	Interior Space - Hard Floors	Hard wood, Laminate, tile	HEPA Vacuum	HEPA Vacuum; Wet Mop	HEPA Vacuum; Wet Mop; Refinish where necessary	HEPA Vacuum; Wet Mop; Refinish; Replace where necessary
3	Interior Space - Walls	Drywall	HEPA Vacuum	HEPA Vacuum; Wet Wipe	HEPA Vacuum; Wet Wipe; Seal & Top Coat	Demo as required; HEPA Vacuum; Wet Wipe, Seal & Top Coat
4	Interior Space - Carpeted Floors	Carpet & Pad	If carpet is G to Excellent and debris loading <20 mg / 100 cm2, vacuum followed by hot water extraction	If carpet is VG to Excellent and Debris loading <20 mg / 100cm2, vacuum followed by hot water extraction	Replace carpet and padding	Replace carpet and padding after demo near end of remodeling
5	Interior Space - Rugs	Rugs	If rug is G to Excellent and Debris loading <20 mg / 100 cm2, vacuum followed by hot water extraction	If rug is VG to Excellent and debris loading <20 mg / 100cm2, vacuum followed by hot water extraction, if odor persists dispose	Dispose of rug	Dispose of rug
6	Interior Space - Curtains, soft blinds	Curtains & soft blinds	If curtains or blinds are in G to Excellent condition, professionally clean	If curtains or blinds are in G to Excellent condition, professionally clean & check odor, if odor persists dispose	Dispose & replace curtains & soft blinds	Dispose & replace Curtains & soft blinds

7	Bedding - Mattress	Mattresses	If mattress are in G to Excellent condition, professionally clean	If mattress are in G to excellent condition, professionally clean	Dispose & replace mattresses & bedding	Dispose & replace after remediation is complete mattresses & bedding
8	Bedding - Pillows, quilts, comforters other padded materials	Pillows, quilts, comforters, other soft bedding materials	If bedding are in G to Excellent condition, professionally clean	Dispose bedding - Pillows, quilts, comforters & replace	Dispose bedding - Pillows, quilts, comforters & replace	Dispose bedding after remediation is complete replace- Pillows, quilts, comforters
8	Clothing, Towels, Sheets, linens other fabric materials that can be washed	Fabric articles that can be laundered	If fabric material are in G to Excellent condition, owner may launder or have professionally cleaned	If fabric material are in G to Excellent condition, owner may launder or have professionally cleaned	If fabric material are in G to Excellent condition, owner may launder or have professionally cleaned - If still odiferous dispose	If fabric material are in G to Excellent condition, owner may launder or have professionally cleaned - If still odiferous or damaged dispose
9	Books & soft toys	High value books, soft & permeable toys	If high value books and permeable goods are in G to Excellent condition, professionally clean if possible or dispose	For high value books and permeable goods that are in G to Excellent condition, professionally clean if possible & deodorize or dispose	Dispose books, soft & permeable toys & replace	Dispose books, soft & permeable toys & replace at the end of remediation
10	Sporting goods, hard	Skies, golf clubs, lanterns, other hard items	HEPA Vacuum	HEPA Vacuum, Wet Wipe	HEPA Vacuum, Wet Wipe	HEPA Vacuum, Wet wipe, Replace where damaged as necessary
11	Sporting goods, soft	Sleeping bags, liners, gloves baklava	If items are in G to Excellent condition, professionally clean - if that is not practical dispose	If items are in G to Excellent condition, professionally clean & check odor, if odor persists dispose	Dispose & Replace	Dispose & Replace
12	Furniture - Upholstered	Upholstered furniture	If upholstered furniture are in G to Excellent condition, professionally clean	If upholstered furniture are in G to Excellent condition, professionally clean, deodorize or dispose	Dispose upholstered furniture & replace	Dispose upholstered furniture & replace after remediation is complete

13	Furniture - Hard, non - permeable	Furniture, Non-permeable	HEPA Vacuum	HEPA Vacuum, Wet Wipe	HEPA Vacuum, Wet Wipe	HEPA Vacuum, Wet Wipe, refinish or if significantly damaged dispose
14	Kitchen - Open & perishable food	Perishable food	Dispose	Dispose	Dispose	Dispose
15	Kitchen - Canned & sealed food	Canned, Sealed food	Wet Wipe & Rinse cans	Wet Wipe & Rinse cans	Dispose	Dispose
16	Kitchen - Cabinets	Kitchen cabinets & shelves	Empty cabinets; HEPA Vacuum clean; Wet Wipe; Deodorize	Empty cabinets; HEPA Vacuum clean; Wet Wipe; Deodorize	Empty cabinets; HEPA Vacuum clean; Wet Wipe; Refinish	Empty cabinets; HEPA Vacuum clean; Wet Wipe; Refinish or Dispose
17	Computer and stereo equipment	High value electronic equipment	HEPA Vacuum, may need professional cleaning if corrosive ash	HEPA Vacuum, may need professional cleaning if corrosive ash	Replace	Replace
18	Kitchen appliances	Refrigerator, other electric appliances	HEPA Vacuum	HEPA Vacuum, may need professional cleaning if corrosive ash	Replace	Replace
19	HVAC	Gas Forced Air	Furnace & ducts professionally cleaned	Furnace & ducts professionally cleaned	Furnace & ducts professionally cleaned - replace if significantly impacted	Replace furnace if significantly impacted
20	Windows	Windows	HEPA Vacuum	HEPA Vacuum, Wet Wipe	HEPA Vacuum, Wet Wipe, replace if PE evaluation stipulates	HEPA Vacuum, Wet Wipe, replace if PE evaluation stipulates
21	Exterior	Siding, decking, roofing & decks	Low pressure wash	Low pressure wash	Low pressure wash, may need PE evaluation for repairs or refinishing	Low pressure wash, will need PE evaluation for repairs or refinishing

References

[1] **Medina, E, ed.** *Technical Guide for Wildfire Impact Assessments for the OEHS Professional.* American Industrial Hygiene Association. 2018.

[2] **King ML.** *Guidelines for Fire and Smoke Damage Repair, Second Edition.* Restoration Industry Association. 2007.

[3] **Portman J.** *After the Fire or Disaster: Dealing with Your Insurance Company.* www.NOLO.com. Published October 10, 2011. Accessed March 8, 2023. https://www.nolo.com/legal-encyclopedia/dealing-with-insurance-after-disaster-29640.html

[4] **Scott SM.** *The Red Guide to Recovery: Resource Handbook for Disaster Survivors; Second Edition.* Heritage Publishing & Communications, Ltd. 2021.

[5] **7 Crucial Steps When Filing a Fire Damage Home Insurance Claim**. Clovered.com. Accessed March 25, 2023. https://clovered.com/does-homeowners-insurance-cover-fire/

[6] **Insurance Information Institute.** Understanding the insurance claims payment process. Insurance Information Institute. Accessed on March 8, 2023. https://www.iii.org/article/understanding-the-insurance-claims-payment-process

[7] **Rocky Mountain Insurance Information Association.** *Colorado - After the Fire: Your Step-by-Step Home Insurance Claim Recovery Guide.* Rocky Mountain Insurance Information. Accessed March 20, 2023. http://www.rmiia.org/downloads/AfterTheFire_StepByStep.pdf

[8] **Bumgardner T.** Guidelines for Reviewing Adjusters' and Contractors' Estimates, *United Policyholders*. Accessed on March 8, 2023. https://uphelp.org/claim-guidance-publications/guidelines-for-reviewing-adjusters-and-contractors-estimates/

[9] **EMSL Analytical, Inc.** Fire and Smoke Damage Testing, Pocket Guide. EMSL Analytical, Inc. https://emsl.com/PDFDocuments/SamplingGuide/Fire_Smoke_Damage_PG_2021.pdf

[10] **Blomqvist P.** *Emissions from Fires Consequences for Human Safety and the Environment.* [Doctoral Thesis (compilation), Division of Fire Safety Engineering]. Department of Fire Safety Engineering and Systems Safety, Lund University. 2005.

[11] **Wentworth GR, Aklilu Y, Landis MS, Hsu Y-M.** Impacts of a large boreal wildfire on ground level atmospheric concentrations of PAHs, VOCs, and ozone. *Atmospheric Environment.* Volume 178. 2018. Pages 19-30. ISSN 1352-2310, https://doi.org/10.1016/j.atmosenv 2018.01.013

[12] **Liao J, Wolfe GM, Hannun RA, et al.** Formaldehyde evolution in US wildfire plumes during the Fire Influence on Regional to Global Environments and Air Quality experiment (FIREX-AQ) *Atmos. Chem. Phys.* 21. 18319–18331. 2021. Accessed on March 8, 2023. https://acp.copernicus.org/articles/21/18319/2021/

[13] **Maliszewska-Kordybach B.** Sources, Concentrations, Fate and Effects of Polycyclic Aromatic Hydrocarbons (PAHs) in the Environment. Part A: PAHs in Air. *Polish Journal of Environmental Studies.* 8. 131-136. 1999.

[14] **Blomqvist P.** *Emissions from Fires Consequences for Human Safety and the Environment.* [Doctoral Thesis (compilation), Division of Fire Safety Engineering]. Department of Fire Safety Engineering and Systems Safety, Lund University. 2005.

[15] **U.S. Environmental Protection Agency**. Method TO-15 Determination of Volatile Organic Compounds (VOCs) In Air Collected in Specially-Prepared Canisters and Analyzed by Gas Chromatography/ ass Spectrometry (GC/MS) *Center for Environmental Research Information. Office of Research and Development*. 1999.

[16] **Baehr C, Smith GJ, Sleeman D, Zevaco TA, Raffelt K, Dahmen N.** Aldehydes and ketones in pyrolysis oil: analytical determination and their role in the aging process. The Royal Society of Chemistry. 2022. doi.org/10.1039/D1RA08899H

[17] **De Vries CJ, Buica A. McKay JBM.** The impact of smoke from vegetation fires on sensory characteristics of Cabernet sauvignon wines made from affected grapes. *South African Journal of Enology and Viticulture*. 37:22-30. 2016.

[18] **Juhlke F, Lorber K, Wagenstaller M, Buettner A.** Influence of the Chemical Structure on Odor Qualities and Odor Thresholds of Halogenated Guaiacol-Derived Odorants. *Front. Chem.* 5:120. 2017. doi: 10.3389/fchem.2017.00120

[19] **Heitmann K, Wichmann H, Bahadir M.** Chemical causes of the typical burnt smell after accidental fires. *Anal Bioanal Chem*. 2009;395(6):1853-1865. doi:10.1007/s00216-009-3071-7

[20] **Heitmann K, Wichmann H, Bahadir M, Gunschera J, Schulz N, Salthammer T.** Chemical composition of burnt smell caused by accidental fires: environmental contaminants. *Chemosphere*. 2011;82(2):237-243. doi:10.1016/j.chemosphere.2010.09.065

[21] **Florian J, Katja L, Maria W, Andrea B.** Influence of the Chemical Structure on Odor Qualities and Odor Thresholds of Halogenated Guaiacol-Derived Odorants. *Frontiers in Chemistry*. 2017. DOI=10.3389/fchem.2017.00120

[22] **Herve E, Price S, Burns G.** Free Guaiacol and 4-Methylguaiacol as Markers of Smoke Taint in Grapes and Wines: Observations from the 2008 Vintage in California. ETS Laboratories. In Œno 2011 Proceedings of the 9th International Symposium of Enology. Dunod Editions, France, 2011. Accessed on March 8, 2023.

[23] **Kjallstrabd J.** Methoxyphenols in Smoke from Biomass Burning. 2000, Thesis for the Degree of Licentiate of Engineering, *Department of Chemical Environmental Sciences, Chalmers University of Technology*. 2000.

[24] **Pitman RM.** Wood ash use in forestry – a review of the environmental impacts. *Forestry: An International Journal of Forest Research*, Volume 79, Issue 5, Pages 563–588. 2006. https://doi.org/10.1093/forestry/cpl041

[25] **American Society for Testing and Materials (ASTM).** *Standard Practice for Sampling and Testing of Possible Carbon Black Fugitive Emissions or Other Environmental Particulate, or Both.* www.astm.org. Accessed March 25, 2023. https://www.astm.org/d6602-13r22.html

[26] **Harper AR., Santin C, Doerr SH, et al.** Chemical composition of wildfire ash produced in contrasting ecosystems and its toxicity to Daphnia magna. *International Journal of Wildland Fire* 28, 726-737. 2019. doi.org/10.1071/WF18200

[27] **Delia A, Baxter D.** The ABCs of Wildfire Residue Contamination Testing, Postfire Assessments of the Indoor Environment. *The Synergist*. 2017.

[28] **Chuang, JC.** Analysis of soil and house dust for polycyclic aromatic hydrocarbons. *U.S. Environmental Protection Agency*, Washington, DC, EPA/600/R-96/060 (NTIS PB96-177712), 1996.

[29] **Wisconsin Department of Health Services.** Polycyclic aromatic hydrocarbons (PAHs). *Wisconsin Department of Health Services.* Accessed March 8, 2023. https://dhs.wisconsin.gov/chemical/pah.htm

[30] **Bradley JN, Magee BH, Allen SI**, Background Levels of Polycyclic Aromatic Hydrocarbons (PAH) and selected Metals in New England Urban Soils, Journal of Soil Contamination, 3(4), 1994

[31] **Maliszewska-Kordybach B.** Sources, Concentrations, Fate and Effects of Polycyclic Aromatic Hydrocarbons (PAHs) in the Environment. Part A: PAHs in Air. *Polish Journal of Environmental Studies.* 8. 131-136. 1999.

[32] **Massachusetts Department of Environmental Protection**. Technical update: background levels of polycyclic aromatic hydrocarbons and metals in soil. *Massachusetts Department of Environmental Protection.* 1992.Accessed on March 8,

2023. https://www.mass.gov/doc/technical-update-background-levels-of-polycyclic-aromatic-hydrocarbons-and-metals-in-soil-0/download

[33] **Concord Scientific Corporation and Beak Consultants.** *Polycyclic Aromatic Hydrocarbons: Sources, Fate and Levels in Air, Water, Soil, Sediments, Sludge and Food in Ontario.* ISBN 0-7729-9281-9. Concord Scientific Corporation and Beak Consultants. 1992.

[34] **US Environmental Protection Agency**. Recommended Toxicity Equivalence Factor (TEF) for Human Health Risk Assessments of 2,3,7,8-Tetrachlorodibenzo-p-dioxin and Dioxin-Like Compounds. U.S. Environmental Protection Agency. 2009. Accessed March 8, 2023.

[35] **Mouat AP, Paton-Walsh C, Simmons JB, Ramirez-Gamboa J, Griffith DWT, Kaiser J.** Emission factors of long-lived volatile organic compounds from the 2019-2020 Australian wildfires during the COALA campaign; discussions. *Atmos. Chem. and Physics.* 2021. https://doi.org/10.5194/acp-2021-742

[36] **Jordan RC, Barrows RR.** *Exterior Wipe Sampling Screening Program*, Apex, NC, Toxics Protection Branch, Air Toxics Analytical Support Team (ATAST) ATAST Response # 06020. North Carolina Department of Environment and Natural Resources. 2006.

[37] **Ruokojärvi P, Aatamila M, Ruuskanen J.** Toxic chlorinated and polyaromatic hydrocarbons in simulated house fires. *Chemosphere*. 2000;41(6):825-828. doi:10.1016/s0045-6535(99)00549-4

[38] **Korobeinichev O, Shmakov A, Paletsky A, et al.** Mechanisms of the Action of Fire-Retardants on Reducing the Flammability of Certain Classes of Polymers and Glass-Reinforced Plastics Based on the Study of Their Combustion. *Polymers*. 2022; 14(21):4523. https://doi.org/10.3390/polym14214523

[39] **World Health Organization**. Dioxins and their effects on human health. Who.int. Published October 4, 2016. https://www.who.int/news-room/fact-sheets/detail/dioxins-and-their-effects-on-human-healthDioxins and their effects on human health https://www.who.int/news-room/fact-sheets/detail/dioxins-and-their-effects-on-human-health

[40] **World Health Organization**. Dioxins and their effects on human health. Who.int. Published October 4, 2016. https://www.who.int/news-room/fact-sheets/detail/dioxins-and-their-effects-on-human-healthDioxins and their effects on human health https://www.who.int/news-room/fact-sheets/detail/dioxins-and-their-effects-on-human-health

[41] **US Centers for Disease Control**. Dioxins, Furans and Dioxin-Like Polychlorinated Biphenyls Factsheet. Published 2019. https://www.cdc.gov/biomonitoring/DioxinLikeChemicals_FactSheet.html

[42] **Simpson SC, Council GW, Talaat J, Covert DC, Teaf CM.** Soil Cleanup Goal for Dioxin Using Probabilistic Risk Assessment Techniques. *Soil and Sediment Contamination: An International Journal*. 25. 00-00. 2006. DOI: 10.1080/15320383.2016.1205549.

[43] **US Environmental Protection Agency**. Recommended Toxicity Equivalence Factor (TEF) for Human Health Risk Assessments of 2,3,7,8-Tetrachlorodibenzo-p-dioxin and Dioxin-Like Compounds. *U.S. Environmental Protection Agency*. 2009. Accessed March 8, 2023.

[44] **World Health Organization**. Dioxins and their effects on human health. Who.int. Published October 4, 2016. https://www.who.int/news-room/fact-sheets/detail/dioxins-and-their-effects-on-human-healthDioxins and their effects on human health https://www.who.int/news-room/fact-sheets/detail/dioxins-and-their-effects-on-human-health

[45] **World Health Organization**. Dioxins and their effects on human health. Who.int. Published October 4, 2016. https://www.who.int/news-room/fact-sheets/detail/dioxins-and-their-effects-on-human-healthDioxins and their effects on human health https://www.who.int/news-room/fact-sheets/detail/dioxins-and-their-effects-on-human-health

[46] **Food and Agriculture Organization/World Health of the United Nations Organization (FAO/WHO)**. JOINT FAO/WHOEXPERT COMMITTEE on FOOD ADDITIVES Sixty-Fourth Meeting.; 2005. Accessed March 25, 2023. https://www.fao.org/3/at877e/at877e.pdf

[47] **Tchounwou PB, Yedjou CG, Patlolla AK, Sutton DJ.** Heavy metal toxicity and the environment. Exp Suppl. 2012;101:133-64. doi: 10.1007/978-3-7643-8340-4_6. PMID: 22945569; PMCID: PMC4144270.

[48] **Tchounwou PB, Yedjou CG, Patlolla AK, Sutton DJ.** Heavy metal toxicity and the environment. Exp Suppl. 2012;101:133-64. doi: 10.1007/978-3-7643-8340-4_6. PMID: 22945569; PMCID: PMC4144270.

[49] **Balali-Mood M, Naseri K, Tahergorabi Z, Khazdair MR, Sadeghi M**. Toxic Mechanisms of Five Heavy Metals: Mercury, Lead, Chromium, Cadmium, and Arsenic. *Front Pharmacol*. 2021 Apr 13;12:643972. doi: 10.3389/fphar.2021.643972. PMID: 33927623; PMCID: PMC8078867.

[50] **Heng YY, Asad I, Coleman B, et al.** Heavy metals and neurodevelopment of children in low and middle-income countries: A systematic review. *PLoS One*. 2022 Mar 31;17(3):e0265536. doi: 10.1371/journal.pone.0265536. PMID: 35358213; PMCID: PMC8970501.

[51] **King ML.** *Guidelines for Fire and Smoke Damage Repair, Second Edition.* Restoration Industry Association; 2007

[52] **US Environmental Protection Agency**. Compendium of methods for determination of Toxic Organic Compounds in Ambient Air, EPA, USEPA/600/4-89/017. 1999. https://www3.epa.gov/ttnamti1/files/ambient/airtox/tocomp99.pdf

[53] **US Environmental Protection Agency**. PA,ORD. IRIS | US EPA. US EPA. Published March 15, 2013. https://www.epa.gov/iris

[54] **Contaminants of Potential Concern Committee of the World Trade Center Indoor Air Task Force Working Group**. World Trade Center Indoor Environment Assessment: Selecting Contaminants of Potential Concern and Setting Health-Based Benchmarks. *Prepared by the Contaminants of Potential Concern (COPC), Committee of the World Trade Center Indoor Air Task Force Working Group.* 2003. Accessed March 8, 2023. https://archive.epa.gov/wtc/web/pdf/contaminants_of_concern_benchmark_study.pdf

[55] **California Department of Toxic Substances Control.** *Persistent and Bioaccumulative Toxic Substances Tables*. Accessed March 25, 2023. https://dtsc.ca.gov/persistent-and-bioaccumulative-toxic-substances/

[56] **Contaminants of Potential Concern Committee of the World Trade Center Indoor Air Task Force Working Group**. World Trade Center Indoor Environment Assessment: Selecting Contaminants of Potential Concern and Setting Health-Based Benchmarks. *Prepared by the Contaminants of Potential Concern (COPC), Committee of the World Trade Center Indoor Air Task Force Working Group.* 2003. Accessed March 8, 2023. https://archive.epa.gov/wtc/web/pdf/contaminants_of_concern_benchmark_study.pdf

[57] **Eurofins Environmental Testing**. A guide to Fire Investigation Testing & Combustion By-Products. eurofins Environmental Testing. Accessed on March 8, 2023. https://cdnmedia.eurofins.com/eurofins-us/media/12160700/cbp-information-packet.pdf

[58] **Baxter DM.** The analysis of fire residue particles using optical microscopy, an overview of method advantages, limitations and ire-related particle terminology. *Environmental Analysis Associates*. Fire analysis methods and terminology-EAA-4 2-24-21, Rev. 2020-4 2-24-21. 20

[59] **EMSL Analytical, Inc.** Fire and Smoke Damage Testing, Pocket Guide. EMSL Analytical, Inc. 2021.

[60] **Eurofins Environmental Testing**. A guide to Fire Investigation Testing & Combustion By-Products. eurofins Environmental Testing. Accessed on March 8, 2023. https://cdnmedia.eurofins.com/eurofins-us/media/12160700/cbp-information-packet.pdf

[61] **Baxter, DM.** The limitations of using the ASTM 6602-13 method for the analysis of wildfire and structure fire residues. *Environmental Analysis Associates*. 2021.Accessed on March 8, 2023.

[62] **Baxter DM**. EAA standard operating procedures: sampling and analysis of surface dust, mold, and combustion residue particles using light microscopy. *Environmental Analysis Associates*. Standard Operating Procedure, Revised March 2013.

[63] **Baxter DM.** The analysis of fire residue particles using optical microscopy, an overview of method advantages, limitations and ire-related particle terminology. *Environmental Analysis Associates*. Fire analysis methods and terminology-EAA-4 2-24-21, Rev. 2020-4 2-24-21. 20

[64] **Baxter DM, Crutcher E, Kovar, B, Larry W.** Analysis of Wildfire and Structure Fire Combustion Residues: Microscopy Methods and Other Considerations. *The Synergist*. August. 28-32. 2022

[65] **Chen H, Samet JM, Bromberg PA, et al**. Cardiovascular health impacts of wildfire smoke exposure. *Part Fibre Toxicol* 18, 2. 2021. https://doi.org/10.1186/s12989-020-00394-8

[66] **US Environmental Protection Agency**. Literature Review of Wipe Sampling Methods for Chemical Warfare Agents and Toxic Industrial Chemicals, Contract No. GS23F0011 L-3 Task Order No. 1125. Prepared for Stephen Billets. 2007.

[67] **Baxter DM, Crutcher R, Kovar B, Wayne L**. Analysis of Wildfire and Structure Fire Combustion Residues: Microscopy Methods and Other Considerations. *The Synergyst*. Nov 2022. https://synergist.aiha.org/202208-fire-combustion-residues

[68] **Brookhaven National Laboratory.** IH75190 Surface wipe sampling for metals. Rev 18 dated 05/10/11. *Brookhaven National Laboratory*. Accessed on March 8, 2023. https://www.bnl.gov/esh/shsd/sop/pdf/ih_sops/ih75190.pdf

[69] **Peacock RD, Cleary TG, Reneke PA, Murphy DC**. A Literature Review of the Effects of Smoke from a Fire on Electrical Equipment (NUREG/CR-7123). National Institute of Standards and Technology. pages 23-24. 2012. **https://www.nrc.gov/reading-rm/doc-collections/nuregs/contract/cr7123/index.html**

[70] **Reagor BT**. Smoke Corrosivity: Generation, Impact, Detection, and Protection. *Journal of Fire Sciences*, 10:169-179, March/April 1992.

[71] **Tewarson A, Chu F**. "Non-Thermal Fire Damage Due to Smoke and Corrosive Compounds. Factory Mutual Research Corporation, Norwood, Massachusetts. 1991.

[72] **US Environmental Protection Agency**. Recommended Toxicity Equivalence Factor (TEF) for Human Health Risk Assessments of 2,3,7,8-Tetrachlorodibenzo-p-dioxin and Dioxin-Like Compounds. U.S. Environmental Protection Agency. 2009. Accessed March 8, 2023. https://www.epa.gov/sites/default/files/2013-09/documents/hhtef_draft_090109.pdf#:~:text=Traditionally%2C%20the%20Toxic%20Equivalency%20Factor%20%28TEF%29%20Methodology%2C%20a,DLCs%20published%20in%202005%20by%20the%20World%20Health

[73] **Larsson F, Andersson P, Blomqvist P, et al**. Toxic fluoride gas emissions from lithium-ion battery fires. *Sci Rep* 7, 10018 (2017). https://doi.org/10.1038/s41598-017-09784-z

[74] **American National Standards Institute/Institute of Inspection Cleaning and Restoration Certification (ANSI/IICRC)**. S500 Standard for Professional Water Damage Restoration: 5th Ed. 2021.

[75] **Baxter DM, Perkins JL, McGhee CR, Seltzer JM.** A regional comparison of mold spore concentrations outdoors and inside "clean" and "mold contaminated" Southern California buildings. *J Occup Environ Hyg*. 2005;2(1):8-18. doi:10.1080/15459620590897523

Author Biography

James Lieberman is certified by the American Board of Industrial Hygiene (AIHA) for the comprehensive practice of industrial hygiene and has over 30 years of experience as a practicing industrial hygienist. He is the founder and leader of Environmental Information Services, Inc, an environmental, health and safety consulting firm based in Boulder, Colorado. He has a B.S. degree in chemistry from the University of Richmond, and an M.B.A. from the University of Colorado.

Mr. Lieberman is the author of *A Practical Guide for Hazardous Waste Management, Administration and Compliance* and *The Business of Environmental Health & Safety Management* and serves as an expert witness in cases involving fire and smoke impacted residences and chemical exposure.

He was honored by Colorado Mountain College in 1984 with an award for Instructor of the Year and in 2009 he was honored by AIHA–Rocky Mountain Section as Industrial Hygienist of the year. He continues to look for ways to engage with and enjoy the communities in his home state of Colorado.

Made in the USA
Monee, IL
13 September 2026

63050002R00077